AMS SHORT COURSE LECTURE NOTES
Introductory Survey Lectures
A Subseries of Proceedings of Symposia in Applied Mathematics

Volume 41 **ROBOTICS**
Edited by R. W. Brockett (Louisville, Kentucky, January 1990)

Volume 40 **MATRIX THEORY AND APPLICATIONS**
Edited by Charles R. Johnson (Phoenix, Arizona, January 1989)

Volume 39 **CHAOS AND FRACTALS: THE MATHEMATICS BEHIND THE COMPUTER GRAPHICS**
Edited by Robert L. Devaney and Linda Keen (Providence, Rhode Island, August 1988)

Volume 38 **COMPUTATIONAL COMPLEXITY THEORY**
Edited by Juris Hartmanis (Atlanta, Georgia, January 1988)

Volume 37 **MOMENTS IN MATHEMATICS**
Edited by Henry J. Landau (San Antonio, Texas, January 1987)

Volume 36 **APPROXIMATION THEORY**
Edited by Carl de Boor (New Orleans, Louisiana, January 1986)

Volume 35 **ACTUARIAL MATHEMATICS**
Edited by Harry H. Panjer (Laramie, Wyoming, August 1985)

Volume 34 **MATHEMATICS OF INFORMATION PROCESSING**
Edited by Michael Anshel and William Gewirtz (Louisville, Kentucky, January 1984)

Volume 33 **FAIR ALLOCATION**
Edited by H. Peyton Young (Anaheim, California, January 1985)

Volume 32 **ENVIRONMENTAL AND NATURAL RESOURCE MATHEMATICS**
Edited by R. W. McKelvey (Eugene, Oregon, August 1984)

Volume 31 **COMPUTER COMMUNICATIONS**
Edited by B. Gopinath (Denver, Colorado, January 1983)

Volume 30 **POPULATION BIOLOGY**
Edited by Simon A. Levin (Albany, New York, August 1983)

Volume 29 **APPLIED CRYPTOLOGY, CRYPTOGRAPHIC PROTOCOLS, AND COMPUTER SECURITY MODELS**
By R. A. DeMillo, G. I. Davida, D. P. Dobkin, M. A. Harrison, and R. J. Lipton (San Francisco, California, January 1981)

Volume 28 **STATISTICAL DATA ANALYSIS**
Edited by R. Gnanadesikan (Toronto, Ontario, August 1982)

Volume 27 **COMPUTED TOMOGRAPHY**
Edited by L. A. Shepp (Cincinnati, Ohio, January 1982)

Volume 26 **THE MATHEMATICS OF NETWORKS**
Edited by S. A. Burr (Pittsburgh, Pennsylvania, August 1981)

Volume 25 **OPERATIONS RESEARCH: MATHEMATICS AND MODELS**
Edited by S. I. Gass (Duluth, Minnesota, August 1979)

Volume 24 **GAME THEORY AND ITS APPLICATIONS**
Edited by W. F. Lucas (Biloxi, Mississippi, January 1979)

Volume 23 **MODERN STATISTICS: METHODS AND APPLICATIONS**
Edited by R. V. Hogg (San Antonio, Texas, January 1980)

Volume 22 **NUMERICAL ANALYSIS**
Edited by G. H. Golub and J. Oliger (Atlanta, Georgia, January 1978)

PROCEEDINGS OF SYMPOSIA IN APPLIED MATHEMATICS

AMS SHORT COURSE LECTURE NOTES

Introductory Survey Lectures

published as a subseries of

Proceedings of Symposia in Applied Mathematics

PROCEEDINGS OF SYMPOSIA
IN APPLIED MATHEMATICS

Volume 41

Robotics

J. Baillieul and D. P. Martin
R. W. Brockett
Bruce R. Donald
Richard M. Murray and S. Shankar Sastry
Madhusudan Raghavan

American Mathematical Society
Providence, Rhode Island

LECTURE NOTES PREPARED FOR THE
AMERICAN MATHEMATICAL SOCIETY SHORT COURSE

ROBOTICS

HELD IN LOUISVILLE, KENTUCKY
JANUARY 16–17, 1990

The AMS Short Course Series is sponsored by the Society's Committee on Employment and Educational Policy (CEEP). The series is under the direction of the Short Course Advisory Subcommittee of CEEP.

Library of Congress Cataloging-in-Publication Data

Robotics/R. W. Brockett, editor; J. Baillieul...[et al.].
p. cm. — (Proceedings of symposia in applied mathematics, ISSN 0160-7634; v. 41)
ISBN 0-8218-0163-5 (acid-free paper)
1. Robotics—Mathematics. I. Brockett, Roger W. II. Baillieul, J. (John) III. American Mathematical Society. IV. Series.
TJ211.M3675 1990 90-1220
629.8′92′0151—dc20 CIP

1980 *Mathematics Subject Classification.* (1985 *Revision*).
Primary 53A17, 68-02, 68G05, 70B15, 93-02.

Printed in the United States of America.
This volume was printed directly from author-prepared copy.
The paper used in this book is acid-free and falls within the guidelines
established to ensure permanence and durability. ∞

10 9 8 7 6 5 4 3 2 1 95 94 93 92 91 90

Table of Contents

Preface

The emergence of the field of robotics has provided the occasion to analyze, and to attempt to replicate, the patterns of movement required to accomplish useful tasks. On the whole, this has been a sobering experience. Just as the ever-closer examination of the physical world occasionally reveals inadequacies in our vocabulary and mathematics, roboticists have found that it is quite awkward to give precise, succinct descriptions of effective movements using the syntax and semantics in common use. Perhaps it has always proved easier to demonstrate than to describe, but in any case, mankind has reached its present state without the benefit of a particularly expressive means for discussing movement. Yet, this is what is needed if we are to convey our wishes to general-purpose robots capable of doing what we ask them to do. In this volume we focus on some of the ways mathematics can be used to address problems in this area.

Because robotics is a broad field, it can be examined with profit from many points of view. The perspectives afforded by computer science, electrical engineering, mechanical engineering, psychology, and neuroscience all yield important insights. Even so, there are pervasive common threads, such as the description of spatial relations and their time evolution. One often finds that ideas from three-dimensional geometry and kinematics are not far from the center of the stage. The concept of a kinematic chain is basic to robotic manipulation, and these objects show up in considerable variety in practical applications. It is, therefore, particularly pleasant to observe that a very natural description of kinematic chains is afforded by a product of one-parameter Lie groups. It turns out that a key step in the design of controllers for industrial robots is equivalent to finding an algorithm for converting between coordinates of the first and second type for the group of rigid motions in three dimensions. We mention a second, perhaps unexpected, mathematical fact related to the manipulation of objects. In considering the application of grasping forces to objects, systems of inequalities play a central role because fingers can only push against, and not pull, objects. In fact, the study of grasping involves convex analysis, models of friction and details of the interface between the hand and the object that go considerably beyond simple mechanics. It happens that many tasks, including, but not limited to, grasping, can be done in more than one way. This may happen because the robot has more than the minimal number of degrees of freedom or because the task description is ambiguous. Disposing of such problems is usually called resolution of redundancy, and in some cases it leads to nonlocal questions in geometry.

From the point of view of the programmer, high-level languages are more efficient than low-level ones. Likewise, in directing the motion of a robot the programmer would like to say as little as possible about the means, preferring to focus attention on the ends. In order to make this possible, it is necessary to incorporate automatic path-planning algorithms in the software and to write compilers that are capable of converting high-level directives into the motor control programs needed to execute motion segments. This means that motion-planning algorithms are an important part of any high-level programming environment and that proving correctness of the motion programs produced by such compilers is an issue.

Computing power is far less expensive now then it has been in the past, and there now exists an effective collection of software tools together with a large group of scientists who know how to use them. The emergence of robotics as a practical activity is one consequence of these developments. The field is immature, and at this stage unification is more easily expressed in terms of the goals than methods. The work presented here demonstrates again the effectiveness of mathematics. Even so, many difficult problems remain.

Roger Brockett

Proceedings of Symposia in Applied Mathematics
Volume **41**, 1990

Some Mathematical Aspects of Robotics[1]

R. W. Brockett

Abstract

We introduce the reader to a few basic problems in robotics, emphasizing kinematic chains and compliance. Using some ideas from differential geometry we give these problems a mathematical interpretation and mention some questions which are partly, but not completely, solved.

1. Introduction

The field of robotics is concerned with the generation of motions which will cause objects in the world to move in some desired way. We may want to generate a motion which will cause a pencil to be sharpened or a motion which will move a silicon wafer from one point to another. The robot itself is ordinarily just a means to an end but, even so, one often views the motion of the robot as being the main object of study. In the literature, the analysis of the associated questions is organized into various sub-problems. Among the well-recognized sub-areas there are:

i) The study of kinematic chains.
ii) The description of compliance.
iii) The problem of grasping.
iv) Motion planning and obstacle avoidance.

Our goal is to introduce some vocabulary and a few problems relating to kinematic chains and compliance control. Figure 1 illustrates the sort of mechanism we are interested in. Although it may not look as though such objects could generate interesting or elegant mathematical questions, we will show that there are, in fact, a rather wide class of mathematical issues which play an important role in this technology.

1980 *Mathematics Subject Classification* (1985 *Revision*). Primary 53A17, 58E20, 70B15, 93–02.
[1]This work was supported in part by the U.S. Army Research Office under grant DAAL03-86-K-0171 and in part by the National Science Foundation under grant CDR-85-00108

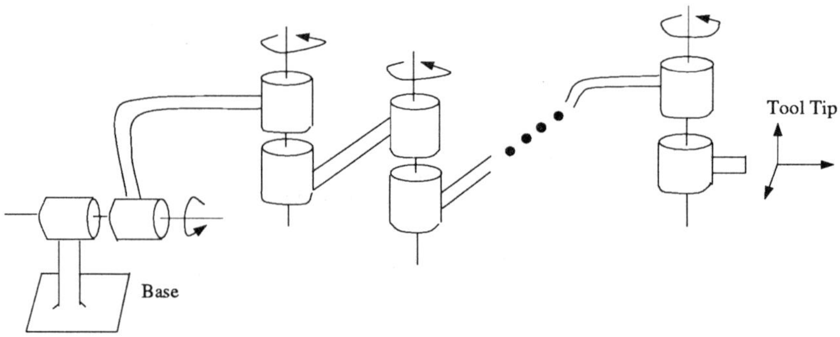

Figure 1. A typical kinematic chain.

2. The Group of Rigid Displacements

We use the notation $\mathbf{E}^3$ to denote ordinary cartesian 3-space with the standard inner product $\langle x, y\rangle = \Sigma x_i y_i$ and refer to it as euclidean 3-space. We let $\mathbf{E}(3)$ denote the set of all length preserving transformations in $\mathbf{E}^3$, i.e. the six-parameter family of translations and (not necessarily proper) rotations. Initially one can think of robots as being used to move rigid objects from one point to another in $\mathbf{E}^3$. For this reason it is not at all surprising that $\mathbf{E}(3)$, or at least the subset $\mathbf{SE}(3)$ which consists of the translations and proper rotations, appears frequently. The groups $\mathbf{E}(3)$ and $\mathbf{SE}(3)$ are called the Euclidean and special Euclidean groups, respectively.

If we describe a particular rigid displacement by the statement that it takes a generic point with coordinate vector x, relative to some fixed coordinate system, to the location having coordinate vector $Ax + b$ relative to the same system, then the effect of composing two such transformations is to send x into $A_2(A_1x + b_1) + b_2 = A_2A_1x + A_2b_1 + b_2$. Recalling the formula from matrix algebra

$$\begin{bmatrix} A_2 & b_2 \\ 0 & 1 \end{bmatrix}\begin{bmatrix} A_1 & b_1 \\ 0 & 1 \end{bmatrix} = \begin{bmatrix} A_2A_1 & A_2b_1 + b_2 \\ 0 & 1 \end{bmatrix}$$

one sees that composition of rigid transformations is captured by a matrix multiplication in the order indicated. It is clear that the composition of two rigid motions is a rigid motion and that each rigid motion has an inverse. Composition is associative so that rigid motions form a group with respect to this operation. Of course A_1 and A_2 must be orthogonal if these transformations are to represent length preserving motions and they must be proper orthogonal matrices (orthogonal with A having a positive determinant) if they are to correspond to physically realizable motions.

We can topologize the space of rigid transformations by introducing a distance measure in terms of this matrix representation. If z is a vector in $\mathbf{E}^n$ let $\|z\|$ denote its euclidean length. If F maps $\mathbf{E}^n$ into $\mathbf{E}^m$ take $\|F\|$ to be the least upper bound on $\|Fz\|/\|z\|$. We may take the distance between T_1 and T_2 to be

$$d(T_1,T_2) = \left\| \begin{bmatrix} A_2 & b_2 \\ 0 & 1 \end{bmatrix} - \begin{bmatrix} A_1 & b_1 \\ 0 & 1 \end{bmatrix} \right\|$$

It is well known and easily checked that, with respect to the corresponding metric topology, the composition law in $\mathbf{SE}(3)$ is continuous. We refer to the above representation as the standard matrix representation of $\mathbf{SE}(3)$. We remark that the group $\mathbf{SE}(3)$ is neither simple nor solvable, but is the semidirect product of the simple group of proper orthogonal transformations ($\mathbf{SO}(3)$) and the three dimensional abelian group consisting of all translations in $\mathbf{E}^3$.

The the path followed by a moving rigid body can be described by associating it with a curve in $\mathbf{SE}(3)$. This means that in a mathematical discussion of manipulation, curves in the group $\mathbf{SE}(3)$ are bound to play a role. Before going into this, however, we recall a few general facts about Lie groups and Lie algebras.

Recall that a *Lie algebra* is a vector space $\mathbf{L}$ with a bilinear operation $[\ ,\ ] : \mathbf{L} \times \mathbf{L} \to \mathbf{L}$ which is antisymmetric and which satisfies the Jacobi identity, $[L_1,[L_2,L_3]] + [L_2[L_3,L_1]] + [L_3,[L_1,L_2]] = 0$. A *Lie group* is a group with a topology on its set of elements such that the group can be given the structure of a differentiable maniford with group multiplication and inversion being continuous. Associated with a Lie group are various one parameter subgroups $g_\alpha(\cdot)$ which satisfy $g_\alpha(t+\sigma) = g_\alpha(t)g_\alpha(\sigma)$. These one parameter subgroups can be identified with elements of a Lie algebra and in this way it is possible to make a one to one correspondence between the elements of the tangent space of the Lie group at the identity and the elements of the Lie algebra. One then uses relations of the type

$$\lim_{t\to 0} g_\alpha(t)g_\beta(t)g_\alpha(-t)g_\beta(-t) = \lim_{t\to 0} g_\gamma(t^2) \quad .$$

to define a Lie bracket $[\alpha,\beta] = \gamma$. (See, e.g. [14]).

For our purposes it is adequate to think in terms of Lie groups and Lie algebras consisting of square matrices. In this case, the one parameter subgroups take the form $M(t) = \exp Lt$ with L being an element of the Lie algebra and exp being defined by $\exp A = I + A + A^2/2! + \ldots$. One sees without difficulty that the Lie algebra of $\mathbf{SE}(3)$ has the representation

$$\mathbf{se}(3) = \left\{ L \mid L = \begin{bmatrix} S & v \\ 0 & 0 \end{bmatrix} \quad ; \quad S = -S^T \right\}$$

We now turn to some properties of the one parameter subgroups of $\mathbf{SE}(3)$ associated with classical theorems of Euler and Chasles. These results can be thought of as asserting the existence of one parameter subgroups which interpolate on $\mathbf{SE}(3)$. We note in passing that the one parameter subgroups of $\mathbf{SE}(3)$ can be visualized as the motion of a threaded screw as it advances in a nut. The pitch of this helical motion, measured in units of length per radian, is clearly independent of the choice of coordinate system, except for the length scale. Going back for more than one hundred years in the literature, the one parameter subgroups of $\mathbf{SE}(3)$ are called *screw motions*. (see [20].)

Lemma 1: For every matrix M of the form

$$M = \begin{bmatrix} A & b \\ 0 & 1 \end{bmatrix} \quad ; \quad A \in \mathbf{SO}(3)$$

a) there exists a matrix N of the form

$$N = \begin{bmatrix} S & x \\ 0 & 0 \end{bmatrix} \quad ; \quad S = -S^T$$

such that $M = \exp Nt|_{t=1}$.

b) there exists a skew symmetric matrix S such that for

$$R(t) = \exp\left(\begin{bmatrix} S & 0 \\ 0 & 0 \end{bmatrix} t\right)\begin{bmatrix} I & A^{-1}b \\ 0 & 1 \end{bmatrix}$$

$M = R(t)|_{t=1}$

c) there exists an orthogonal matrix Ψ and a vector d such that

$$\begin{bmatrix} A & b \\ 0 & 1 \end{bmatrix} = \begin{bmatrix} \Psi & d \\ 0 & 1 \end{bmatrix}\begin{bmatrix} A_0 & b_0 \\ 0 & 1 \end{bmatrix}\begin{bmatrix} \Psi^T & -\Psi^T d \\ 0 & 1 \end{bmatrix}$$

with A_o and b_0 of the form

$$A_o = \begin{bmatrix} \cos\theta & \sin\theta & 0 \\ -\sin\theta & \cos\theta & 0 \\ 0 & 0 & 1 \end{bmatrix} \quad ; \quad b_o = \begin{bmatrix} 0 \\ 0 \\ e \end{bmatrix}$$

The proof of this lemma offers no difficulty but we do want to make explicit contact with geometry. Part b is a version of Euler's theorem which states that any rigid motion can be thought of as being the result of a translation followed by a rotation. Chasles is usually credited with the statement that any rigid displacement can be achieved by a rotation and a translation which commute. Since it is easy to see that a rotation about the origin, represented by

$$R = \begin{bmatrix} A & 0 \\ 0 & 1 \end{bmatrix}$$

and a translation, represented by

$$T = \begin{bmatrix} I & b \\ 0 & 1 \end{bmatrix}$$

commute if, and only if, $Ab = b$ we can see that part c is an expression of the result of Chasles. With respect to part c, notice that a_o and b_o are unique and that unless A is the identity, we can make the choice of d unique by imposing the added condition $\langle d, b_o \rangle = 0$.

Inevitably, a robotic device will fail to able to perform all possible translations and rotations. Different choices at the design stage will lead to different limitations and one is led to ask about the extent to which one can make meaningful comparisons between the capabilities of different designs. This calls for an examination of certain questions involving the definition of length and volume in the space of euclidean motions and the remainder of this section will be devoted to this.

The double bracket expression $[L_1, [L_2, \cdot]]$ defines a mapping the Lie algebra into itself. Since it is a linear map of a vector space into itself it has a trace. Thought of as a function of L_1 and L_2, the trace defines a bilinear form on $\mathbf{L} \times \mathbf{L}$. This form is symmetric by virtue of the identity $tr(T_1T_2) = tr(T_2T_1)$. It is called the *Killing form*; we denote by $\langle \ , \ \rangle_K$. (See [17] for a leisurely discussion of this idea.) Being bilinear and symmetric, the Killing form defines a quadratic form on the Lie algebra and hence on the tangent space of a Lie group. This form need not be definite or even nondegenerate; it can be identically zero. However, Weyl showed that it is negative definite for a large and important class of Lie groups, namely, the compact semisimple ones. In the case of the orthogonal group acting on $\mathbf{E}^3$, the Lie algebra can be thought of as the set of three by three skew symmetric matrices. One verifies easily that, in this case, the Killing form is just $-2(w_{12}^2 + w_{13}^2 + w_{23}^2)$.

If the Killing form is negative definite, it can be used to define a Riemannian metric on the group. We will work this out in the special case of $\mathbf{SO}(3)$. Introducing Euler angles and doing some explicit computation we show how to translate the Killing form at the tangent plane associated with the identity to the tangent plane at any other point on the Lie group. For the parameterization of $\mathbf{SO}(3)$ via

$$\Theta = \begin{bmatrix} \cos\theta & \sin\theta & 0 \\ -\sin\theta & \cos\theta & 0 \\ 0 & 0 & 1 \end{bmatrix} \begin{bmatrix} \cos\phi & 0 & \sin\phi \\ 0 & 1 & 0 \\ -\sin\phi & 0 & \cos\phi \end{bmatrix} \begin{bmatrix} \cos\psi & \sin\psi & 0 \\ -\sin\psi & \cos\psi & 0 \\ 0 & 0 & 1 \end{bmatrix}$$

which covers the set of orthogonal matrices with positive determinant twice as θ, ϕ, and ψ vary over $[0, 2\pi]$, the Riemannian metric that comes from translating - $1/2 \ \langle \ , \rangle_K$ is just - $1/2 \ \langle \ d\Theta\Theta^T, d\Theta\Theta^T \rangle_K$ so that

$$(ds)^2 = (d\theta)^2 + (d\phi)^2 - 2\sin\phi(d\psi)(d\theta) + (d\psi)^2$$

(We use a factor of $1/2$ so that the measure of the circle $0 \leq \psi \leq 2\pi$ will be 2π.)
If we write this in terms of a 3 by 3 symmetric matrix

$$(ds)^2 = (d\theta, d\phi, d\psi)G(d\theta, d\phi, d\psi)^T$$

then $\det G = \cos^2\phi$. Since the volume of a Riemannian manifold is the integral of $(\det G)^{1/2}$ and since integrating over $\mathbf{SO}(3)$ twice gives

$$\int_0^{2\pi}\int_0^{2\pi}\int_0^{2\pi} (\det G)^{1/2} d\theta d\phi d\psi = 16\pi^2$$

we see that the volume of $\mathbf{SO}(3)$ is $8\pi^2$. (Some authors use a factor of $1/4$ in the formula for $(ds)^2$, and thus get a volume form which differs by a factor of 1/8.)

On **se**(3) the Killing form is the trace of T where T is the linear map which acts according to

$$T\left(\begin{bmatrix} A & a \\ 0 & 0 \end{bmatrix}\right) = \left[\begin{bmatrix} \Omega_1 & v_1 \\ 0 & 0 \end{bmatrix}, \left[\begin{bmatrix} \Omega_2 & v_2 \\ 0 & 0 \end{bmatrix}, \begin{bmatrix} A & a \\ 0 & 0 \end{bmatrix}\right]\right]$$

This works out to yield $(tr\Omega_1\Omega_2)$ and since v_1 and v_2 play no role this form is clearly degenerate on **se**(3). Since $-tr(\Omega_1\Omega_1)$ is twice the sum of the squares of the entries of the skew-symmetric matrix Ω_1 we see that the Killing form on **se**(3) has rank 3. Because it is degenerate it does not define a Riemannian metric on **SE**(3). In fact, there is no canonical choice of Riemannian metric on **SE**(3).

If $G = \{M\}$ is a set of n by n matrices which form a Lie group under matrix multiplication, and if L is an element of the corresponding Lie algebra, then we can think of LM as defining a "right invariant" vector field on G. That is to say, just as we saw that L can be interpreted as an element of the tangent space at the identity, LM belongs to the tangent space at M and is the "translation" of L by multiplication by M on the right. In an analogous way, we can think of ML as defining a left invariant vector field on G.

If S is a subset of **SE**(3) and M is an element of **SE**(3) then SM is the right translation of S. A volume measure on **SE**(3) could not be thought of as being natural unless it assigned the same measure to S and SM and, for that matter, MS. There is just one (up to choice of scale) natural volume measure on **SE**(3). This is the product of the volume measure on the rotational part (as defined above) and the standard euclidean volume measure on the translational part. In terms of the parameterization of **SE**(3) given by

$$M = \begin{bmatrix} \Theta & x \\ 0 & 1 \end{bmatrix}$$

with $\Theta \in \mathbf{SO}(3)$ and parametized as above, this volume form is (up to a scale factor)

$$Vol(S) = \int_s |cos\phi| d\theta d\phi d\psi dx_1 dx_2 dx_3$$

In addition to this volume form, there is a natural (up to choice of scale again) bilinear form on **se**(3) which, in combination with the Killing form, measures the pitch of the screw motion exp(Lt). The idea is this. If

$$L = \begin{bmatrix} \Omega & v \\ 0 & 0 \end{bmatrix}$$

with

$$\Omega = \begin{bmatrix} 0 & w_3 & -w_2 \\ -w_3 & 0 & w_1 \\ w_2 & -w_1 & 0 \end{bmatrix}$$

and

$$v = \begin{bmatrix} v_1 \\ v_2 \\ v_3 \end{bmatrix}$$

then

$$\alpha = (w_1v_1 + w_2v_2 + w_3v_3)/(w_1^2 + w_2^2 + w_3^2)$$

is the pitch of the screw motion exp(Lt). It is expressed in units of length per radian. Since the denominator in this expression just one half the Killing form, and hence invariant, and since the ratio is clearly invariant, the numerator must be invariant as well. Fix a choice of length scale and define a map as follows:

$$* : \mathbf{se}(3) \to \mathbf{se}(3)$$

such that

$$* \begin{bmatrix} \Omega & v \\ 0 & 0 \end{bmatrix} = \begin{bmatrix} *v & *\Omega \\ 0 & 0 \end{bmatrix}$$

with

$$*\Omega = \begin{bmatrix} \omega_1 \\ \omega_2 \\ \omega_3 \end{bmatrix}$$

and

$$*v = \begin{bmatrix} 0 & v_3 & v_2 \\ -v_3 & 0 & v_1 \\ v_2 & -v_1 & 0 \end{bmatrix}$$

Then

$$2(v_1\omega_1 + v_2\omega_2 + v_3\omega_3) = -1/2\left(\langle L, *L\rangle_K + \langle *L, L\rangle_K\right)$$

Altering the choice of length scale simply scales this quantity and thus we see that the form $1/2\left(\langle L, *L\rangle_K + \langle *L, L\rangle_K\right)$ is, up to a scale factor, a naturally defined bilinear form on $\mathbf{se}(3)$. It is sometimes called the Klein form; the mapping it defines from L to its dual is closely related to the Hodge star operator. We will have some additional use for this when describing forces and torques in section 7.

3. Euclidean Graphs

By a *euclidean graph* G in $\mathbf{E}^3$ we understand a set of vertices and a set of edges $G = (V, E)$ with V and E having the following three properties. *a)* The graph is connected and its branches are directed. *b)* Associated with each of the vertices is a point in $\mathbf{E}^3$ and an orthonormal frame. Exactly one vertex, the root vertex, corresponds to $0 \subset \mathbf{E}^3$ and the orthonormal frame represented by the identity matrix. *c)* Define the $\mathbf{SE}(3)$ element

$$T_{0i} = \begin{bmatrix} \Theta^{0i} & b^{0i} \\ 0 & 1 \end{bmatrix}$$

where b^{0i} is the vector from the root vertex, to vertex i, expressed in the coordinate system associated with the root node and where the columns of Θ^{0i} are the unit vectors of the ordered basis at the ith vertex, expressed in the coordinate system of the root vertex. If there is a branch from node i to node j, it is labeled with

$$T_{ij} = \begin{bmatrix} \Theta^{ij} & b^{ij} \\ 0 & 1 \end{bmatrix} = T_{0i}^{-1} T_{0j}$$

Of course we could also describe T_{ij} by saying that b^{ij} is the vector from i to j expressed in the coordinate system associated with the frame at node i and that the columns of Θ^{ij} are the unit vectors of the frame at j expressed in terms of the coordinate system at i. Figure 2 illustrates this idea. The basic cycle condition, or consistency relation, is

$$T_{ij} T_{jk} T_{ki} = I$$

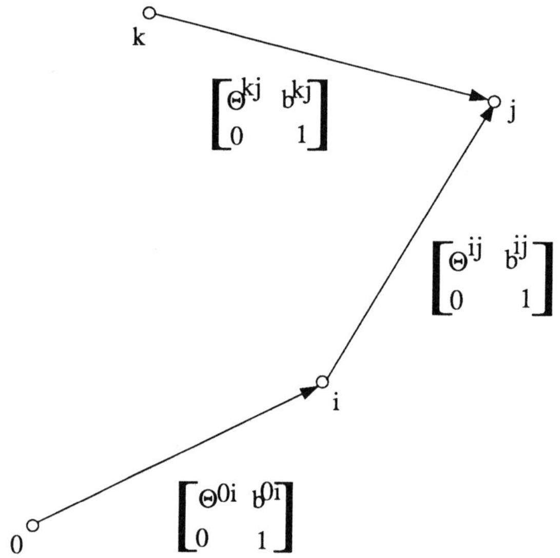

Figure 2: Defining euclidean graphs

This data structure lets us keep track of spatial relations. Graph theory plays a key role in the formulation of the equations describing the forces and displacements in structures and the voltages and currents in electrical networks; many of the standard ideas from this type of analysis are applicable here. For example, the idea of a maximal tree, a basis of loops, duality, etc. show up in the analysis of specific questions. If the euclidean graph is associated to objects which deform over time then some of the T_{ij} which appear will depend on parameters. We will make use of such parametrized euclidean graphs in the next section on kinematic chains and also in a later section on map making.

4. Kinematic Chains

To a first approximation robotic manipulators consist of rigid links joined together at single degree of freedom joints (see figure 1). The systematic description of robotic kinematics leads to the study of kinematic chains and, as we are about to show, the one parameter subgroups of $\mathbf{SE}(3)$ play a basic role in this subject.

Consider a rigid link together with a line in space about which this link can rotate. Locate one coordinate system at the intersection of this line and the link. Suppose that a second coordinate system is fixed in the link. As the link rotates about the line, the coordinate system on the link moves relative to the one on the line in a very simple way. If we fix a right-handed triad of orthogonal vectors at the tip of each member of the chain, it is not too difficult to see that the euclidean transformation, which describes the position and orientation of the $(i+1)^{\text{st}}$ triad in terms of the ith, is

$$\begin{bmatrix} M_i & b_i \\ 0 & 1 \end{bmatrix} \exp \begin{bmatrix} S_i & 0 \\ 0 & 0 \end{bmatrix} \theta_i$$

Thus, the triad fixed at the free end is related to that at the base by the product

$$T(\theta_1, \theta_2, \ldots, \theta_r) = M_1 \left(\exp \theta_1 \begin{bmatrix} S_r & 0 \\ 0 & 0 \end{bmatrix} \right) M_2 \left(\exp \theta_2 \begin{bmatrix} S_2 & 0 \\ 0 & 0 \end{bmatrix} \right) \ldots M_r \left(\exp \theta_r \begin{bmatrix} S_1 & 0 \\ 0 & 0 \end{bmatrix} \right)$$

where the M_i is the element of the euclidean group which maps the coordinate system at the end where it joins the $(i-1)^{\text{st}}$ element to the coordinate system where it joins the $(i+1)^{\text{st}}$.

It is easy to see that $P(\exp M)^{-1}P^{-1} = \exp(PMP^{-1})$. Thus we can use the identity $M \exp N = (\exp MNM^{-1})M$ repeatedly to write

$$T(\theta_1, \theta_2, \ldots, \theta_r) = Me^{H_1\theta_1}e^{H_2\theta_2} \ldots e^{H_r\theta_r}$$

Example: To fix ideas consider the two degrees of freedom system shown in Figure 3. There are four vertices in the corresponding euclidean graph. Two of the transformations (T_{ij}) are constant and characteristic of the links; Two are θ-dependent. The transformations

$$T_{04} = \begin{bmatrix} \cos\theta_1 & \sin\theta_1 & 0 & 0 \\ -\sin\theta_1 & \cos\theta_1 & 0 & 0 \\ 0 & 0 & 1 & 0 \\ 0 & 0 & 0 & 1 \end{bmatrix} \begin{bmatrix} 1 & 0 & 0 & 0 \\ 0 & 1 & 0 & \ell_1 \\ 0 & 0 & 1 & \ell_2 2 \\ 0 & 0 & 0 & 1 \end{bmatrix} \begin{bmatrix} \cos\theta_2 & 0 & \sin\theta_2 & 0 \\ 0 & 1 & 0 & 0 \\ -\sin\theta_2 & 0 & \cos\theta_2 & 0 \\ 0 & 0 & 0 & 1 \end{bmatrix} \begin{bmatrix} 1 & 0 & 0 & 0 \\ 0 & 1 & 0 & 0 \\ 0 & 0 & 1 & \ell_3 \\ 0 & 0 & 0 & 1 \end{bmatrix}$$

or

$$\begin{aligned} T_{04} &= e^{\Omega_1\theta_1} M_1 e^{\Omega_2\theta_2} M_2 \\ &= \tilde{M} e^{\tilde{\Omega}_1\theta_1} e^{\tilde{\Omega}_2\theta_2} \end{aligned}$$

where $\tilde{\Omega}_2 = M_2^{-1}\Omega_2 M_2, \tilde{\Omega}_1 = M_2^{-1}M_1^{-1}\Omega_1 M_1 M_2$ and $\tilde{M} = M_1 M_2$

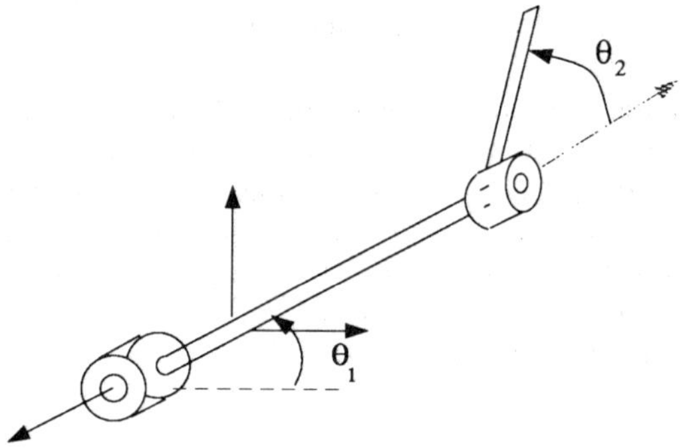

Figure 3: A two degree of freedom mechanism.

This product of exponentials formula not only applies to chains of the form shown in figures 1 and 3, but also applies to mechanisms containing joints which allow translational motion only. These are usually called prismatic joints. In this case we must allow the product to include generators of the translation elements, i.e., factors of the form

$$\exp\begin{bmatrix} 0 & b_i \\ 0 & 0 \end{bmatrix}\theta_i$$

It is to be emphasized that any kinematic chain defines an M and the $H_1, H_2, \ldots, H_r$ with the H_i being elements of the Lie algebra of $\mathbf{se}(3)$ and the M and element of the Lie groups $\mathbf{E}(3)$. Once a choice for the coordinate systems at the beginning and end of the chain is made and a definition of which angles correspond to $\theta_i = 0$ is selected, those matrices characterize, and are characterized by, the chain.

As an example of the type of kinematic chain which is of special interest in practical robotics we mention the type of chain which arises in the description of robotic wrists. The simplest wrist is a three axis kinematic chain in which the three axes of rotation all intersect at a single point. This means that there are no translational terms and that

$$T_{03}(\theta_1, \theta_2, \theta_3) = e^{A_1\theta_1} e^{A_2\theta_2} e^{A_3\theta_3}$$

with A_1, A_2, A_3 all being of the form

$$A_i = \begin{bmatrix} \Omega_i & 0 \\ 0 & 0 \end{bmatrix}$$

The easiest case to analyze is the case where Ω_1 and Ω_2 are orthogonal and Ω_2 and Ω_3 are orthogonal in the sense that they represent rotations about axes which are orthogonal. (This is equivalent to the statements $\langle\Omega_1, \Omega_2\rangle_K = \langle\Omega_2, \Omega_3\rangle_K = 0$). The Euler angles discussed in

Section 2 could be interpreted as describing such a situation. If this condition is violated, as it is in the design of the wrist used by one large manufacturer, then $T_{03}(\theta_1, \theta_2, \theta_3)$ does not map onto $S0(3)$ but only onto a portion of it. This fraction of the volume covered is $\cos\alpha_1 \cos\alpha_2$ where α_1 and α_2 are the angles between axis 1 and axis 2, and axis 2 and axis 3, respectively.

We would like to emphasize that insofar as the expressions of the kinematics of chains without loops, the product of exponentials completely eliminates the need for the rather tedious display of formulas which figure so prominently in the literature.

5. Inverse Kinematics

An easy calculation shows that

$$\exp\begin{bmatrix} 0 & \theta & 0 & 0 \\ -\theta & 0 & 0 & 0 \\ 0 & 0 & 0 & \alpha\theta \\ 0 & 0 & 0 & 0 \end{bmatrix} = \begin{bmatrix} \cos\theta & \sin\theta & 0 & 0 \\ -\sin\theta & \cos\theta & 0 & 0 \\ 0 & 0 & 1 & \alpha\theta \\ 0 & 0 & 0 & 1 \end{bmatrix}$$

Unless α is zero the one parameter family $(\cos\theta, \sin\theta, \alpha\theta)$ in $\mathbf{R}^3$ is not the zero locus of a set of algebraic equations. Thus it is not possible to eliminate transcendental functions from the study of screw motions. However, if the elements of the kinematic chain are all pure rotation or pure translation, it is possible to reexpress matters in such a way as to eliminate transcendental functions. Thus, we can algebraically parameterize the one parameter subgroup associated with pure rotations

$$\{G|G = \exp\begin{bmatrix} 0 & \theta & 0 & 0 \\ -\theta & 0 & 0 & 0 \\ 0 & 0 & 0 & 0 \\ 0 & 0 & 0 & 0 \end{bmatrix} ; -\infty < \theta < \infty\} = \{G|G = \begin{bmatrix} x & y & 0 & 0 \\ -y & x & 0 & 0 \\ 0 & 0 & 1 & 0 \\ 0 & 0 & 0 & 1 \end{bmatrix} ; x^2 + y^2 = 1\}$$

and obviously we may describe the one parameter subgroups associated with translation in purely algebraic terms.

The task which a robot is to perform is usually described in world coordinates, whereas it is the θ's which are controlled by the control system of the robot. For this reason, a basic problem in robotics is that of solving the equation in $\mathbf{SE}(3)$

$$T = M \exp H_1\theta_1 \ldots \exp H_6\theta_6$$

for $\theta_1, \theta_2, \ldots, \theta_6$. This is called the inverse kinematics problem. M. Raghavan's notes discuss it in some detail. The inverse kinematics problem is a special case of the following more general problem.

Let $\mathbf{G}$ be a k-parameter family of n by n matrices which form a Lie group. Let $\mathbf{L}$ be the corresponding Lie algebra and let $L_1, L_2, \ldots, L_k$ an ordered basis for the algebra. One says that

$$T_1(v) = e^{L_1v_1+L_2v_2+\cdots+L_kv_k}$$

and

$$T_2(\theta) = e^{L_1\theta_1}e^{L_2\theta_2}\dots e^{L_k\theta_k}$$

define coordinates for $\mathbf{G}$ of the first and second kind, respectively. If we assume that the $\mathbf{G}$ is an algebraic group, in the sense that as a subset of $Gl(n)$ it can be characterized as the locus of solutions of a finite set of polynomial equations,

$$\mathbf{G} = \{G|\ \psi_i(g_{11}, g_{12}, \dots g_{nn}) = 0;\ i = 1, 2, \dots r\}$$

and if we assume that each of the one parameter groups $e^{L_i\theta_i}$ corresponding to the given basis of the Lie algebra defines a set in $\mathbf{G}$ which is also the locus of solutions of a finite set of algebraic equations, then $T_2(\theta) = G$ is equivalent to a set of algebraic equations. To be more specific, suppose that $e^{L_i\theta_i}$ can be written as

$$\left\{e^{L_i\theta}| - \infty < \theta < \infty\right\} = \{G|\phi_{ik}(g_{11}, g_{12}, \dots g_{nn}) = 0\ ;\ k = 1, 3\dots\ell_i\}\ .$$

with ϕ_{ij} being polynomials. Let $F(g_{11}, g_{12}, \dots, g_{nn})$ denote the field of rational functions in $g_{11}, g_{12}, \dots, g_{nn}$. A question of general interest is that of determining how the degree of these equations, and Galois group associated with the extension of this field obtained by adding to $F(g_{11}, g_{12}, \dots g_{nn})$ the roots of these equations, depend on $\mathbf{G}$ and the ordered basis $L_1, L_2, \dots, L_k$.

Example: If the group is the group of 3 by 3 orthogonal matrices and if we let L_1, L_2, L_3 be given by

$$L_1 = \begin{bmatrix} 0 & 1 & 0 \\ -1 & 0 & 0 \\ 0 & 0 & 0 \end{bmatrix}\ ;\ L_2 = \begin{bmatrix} 0 & 0 & 1 \\ 0 & 0 & 0 \\ -1 & 0 & 0 \end{bmatrix}\ ;\ L_3 = \begin{bmatrix} 0 & 0 & 0 \\ 0 & 0 & 1 \\ 0 & -1 & 0 \end{bmatrix}$$

then the equations described above are 21 in number and take the form

$$\begin{bmatrix} a & b & 0 \\ -b & a & 0 \\ 0 & 0 & 1 \end{bmatrix}\begin{bmatrix} c & 0 & d \\ 0 & 1 & 0 \\ -d & 0 & c \end{bmatrix}\begin{bmatrix} 1 & 0 & 0 \\ 0 & e & f \\ 0 & -f & e \end{bmatrix} = \begin{bmatrix} g_{11} & g_{12} & g_{13} \\ g_{21} & g_{22} & g_{23} \\ g_{31} & g_{32} & g_{33} \end{bmatrix}$$

$$a^2 + b^2 = c^2 + d^2 = e^2 + f^2 = 1$$

$$G^TG = I$$

In this case the total degree of the system is 8 and the Galois group is $\mathbf{Z}_2 \times \mathbf{Z}_2 \times \mathbf{Z}_2$. It is not difficult to write down an explicit solution for the canonical coordinates of the second kind involving only rational operations and square roots.

6. Velocities

If $R(t)$ is a map of $[0, t_1]$ into $\mathbf{E(3)}$ then we can express $\dot{R}(t)$ as

$$\dot{R}(t) = \begin{bmatrix} \Omega & v \\ 0 & 0 \end{bmatrix} R(t)$$

in which case Ω and v represent the velocity in the fixed coordinate system, or as

$$\dot{R}(t) = R(t)\begin{bmatrix} \Omega & v \\ 0 & 1 \end{bmatrix}$$

in which case Ω and v are the velocities expressed in the moving coordinate system. In either case the velocities associated with rigid motion are identified with the elements of the Lie algebra of $\mathbf{SE}(3)$.

On the other hand, in mechanics books one sees velocity represented by pairs of vectors (v, w) with v being the rectilinear velocity and w being the angular velocity. In order to discuss the relationship between these two different points of view it is convenient to recall the operator $*$ defined in section 2 by the formulae

$$*\begin{bmatrix} 0 & w_3 & -w_2 \\ -w_3 & 0 & w_1 \\ w_2 & -w_1 & 0 \end{bmatrix} = \begin{bmatrix} w_1 \\ w_2 \\ w_3 \end{bmatrix}$$

$$*\begin{bmatrix} w_1 \\ w_2 \\ w_3 \end{bmatrix} = \begin{bmatrix} 0 & w_3 & -w_2 \\ -w_3 & 0 & w_1 \\ w_2 & -w_1 & 0 \end{bmatrix}.$$

A key identity in involving $*$ is $\Theta(*\Omega) = *(\Theta\Omega\Theta^T)$ or, equivalently, $*(\Theta w) = \Theta(*w)\Theta^T$ whose verification we leave to the reader.

Under a change of coordinates an element of $\mathbf{se}(3)$ transforms according to

$$\begin{bmatrix} \Theta & b \\ 0 & 1 \end{bmatrix}\begin{bmatrix} \Omega & v \\ 0 & 0 \end{bmatrix}\begin{bmatrix} \Theta^T & -\Theta^T b \\ 0 & 1 \end{bmatrix} = \begin{bmatrix} \Theta\Omega\Theta^T & -\Theta\Omega\Theta^T b + \Theta v \\ 0 & 0 \end{bmatrix}$$

This is an example of the *adjoint action* of a Lie group on its Lie algebra. It is commonly written as $Ad_g(L)$. (See [17].) If we express this transformation in terms of v and $w = *\Omega$ it becomes

$$\begin{bmatrix} v' \\ w' \end{bmatrix} = \begin{bmatrix} \Theta & \Theta(*b)\Theta^T \\ 0 & \Theta \end{bmatrix}\begin{bmatrix} v \\ w \end{bmatrix} = \begin{bmatrix} \Theta v + \Theta(*b)\Theta^T w \\ \Theta w \end{bmatrix}$$

When $\Theta = I$ this corresponds to the usual cross product rule found in mechanics books $v \to v + w \times b$. This representation of $\mathbf{SE}(3)$ in terms of six by six matrices is sometimes more useful then the standard four by four representation introduced previously.

7. Forces and Torques

In the previous sections we have emphasized the group of rigid displacements and its associated Lie algebra, thought of as the set of velocity vectors associated with curves in $\mathbf{SE}(3)$ which are parametized by time. This was facilitated by the use of a certain matrix representation for this group and, by this choice, a certain matrix representation for the Lie algebra. Everyday experience, however, makes it clear that it is a combination of velocities and forces, or torques, which make things happen, and therefore that we need to see how forces and torques can best be thought of in this framework.

Work is the time-integral of the product of force and velocity, or torque, and angular velocity. This suggests that force/torque should be thought of as belonging to a space which is the linear algebraic dual of the velocity space.

As we have seen earlier, if we have a smoothly evolving rigid motion T(t) we can choose to describe its velocity by choosing a point in $\mathbf{E}^3$ and ordered orthonormal basis at that point; considering this as the identity element of $\mathbf{SE}(3)$, find

$$N(t) = \begin{bmatrix} \Omega(t) & v(t) \\ 0 & 0 \end{bmatrix}$$

with $\Omega = \Omega^T$ being the skew-symmetric matrix representation of the angular velocity and v(t) $\in$ E^3 the linear velocity. If we change the choice of point and choice of ordered orthonomal basis then N becomes

$$N'(t) = \begin{bmatrix} \Omega'(t) & v'(t) \\ 0 & 0 \end{bmatrix}$$

$$= \begin{bmatrix} \Theta & b \\ 0 & 1 \end{bmatrix} \begin{bmatrix} \Omega(t) & v(t) \\ 0 & 0 \end{bmatrix} \begin{bmatrix} \Theta^T & -\Theta^T b \\ 0 & 1 \end{bmatrix}$$

Consider a coordinate system moving with a rigid body. If we have a force/torque pair (F,τ) applied at the origin of this system with (F,τ) being represented as a skew-symmetric matrix and a 3-vector, respectively, i.e.

$$F(t) = \begin{bmatrix} 0 & f_3(t) & -f_2(t) \\ -f_3(t) & 0 & f_1(t) \\ f_2(t) & -f_1(t) & 0 \end{bmatrix}$$

and
$$\tau = \begin{bmatrix} \tau_1 \\ \tau_2 \\ \tau_3 \end{bmatrix}$$

then the change of coordinates formula works like this. Changing the point of application of the force and rotating the coordinate frame that we use to describe F yields.

$$\begin{bmatrix} F'(t) & \tau'(t) \\ 0 & 0 \end{bmatrix} = \begin{bmatrix} \Theta^T & -\Theta^T b \\ 0 & 1 \end{bmatrix} \begin{bmatrix} F(t) & \tau(t) \\ 0 & 0 \end{bmatrix} \begin{bmatrix} \Theta & b \\ 0 & 1 \end{bmatrix}$$

On the other hand, in order to compute the work done as a rigid body moves along with path one must pair torque with angular velocity and force with linear velocity. One can express the matter in terms of the * operator introduced in section 2.

$$work = \int_o^{t1} 1/2 \left(\langle N(t), *\tilde{N}(t) \rangle_K + \langle *N(t), \tilde{N}(t) \rangle_K \right) dt$$

where $\tilde{N}$(t) is the force by force/torque matrix displayed above. As in the definition of the Klein form, there is an implicit use of a length scale here.

The reference to "free" versus "bound" vectors which one sees in books on statics and mechanics can be said to stem from the triangular nature of these transformations. Since Ω' does not depend on b, the angular velocity Ω is said to be a free vector; likewise, F' does not depend on b so force is also said to be a free vector. Linear velocity and torque are called bound vectors.

In one dimension the relationship between force and displacement is often quantified by a spring constant as in the familiar equation $f = kx$. Controlling the "spring constant" associated with various motions such as shaking hands and assembling a snap-together object is a critical part of motion control. In the robotics literature this is called *compliance control* and it is one of the more deeply studied aspects of the subject. [6-10]

From the above remarks we can see that for rigid motions in $\mathbf{E}^3$ the linear maps from velocities (linear and rotational) to forces/torques are maps from a six dimensional Lie algebra into its dual. If we assume, and this is certainly an important case, that the spring forces are derivable from a potential energy, then after a choice of coordinates the linearization of this map takes the form

$$\begin{bmatrix} f \\ T \end{bmatrix} = \begin{bmatrix} S & R \\ R^t & Q \end{bmatrix} \begin{bmatrix} \delta x \\ \delta\Theta \end{bmatrix}$$

with S and Q being 3 by 3 symmetric matrices. It is this data which specifies the (linearized) compliance.

8. Uniform Coverage and Harmonic Maps

From section two we see that joint angles can be thought of as a specific choice of canonical coordinates of the second kind for the Lie group of rigid motions. From section seven we see that forces/torque can be thought of as a linear functional on the tangent space of the group of rigid motions. If we restrict our attention to the case where we have a six degree of freedom kinematic claim with all joints being revolute, then we have a map of the six torus T^6 into $\mathbf{E}(3)$ given by

$$f\ :\ (\theta_1, \theta_2, \ldots, \theta_6) \mapsto Me^{A_1\theta_1}e^{A_2\theta_2}\ldots e^{A_6\theta_6}$$

This map has a linearization df. Dual to the $\dot{\theta}$'s are a set of torques $\tau_1, \tau_2, \ldots, \tau_6$ and dual to the velocity in $\mathbf{E}(3)$ we have a set of force-torques. These ideas are connected in the following important but elementary way. We can compute the rate of work done by the forces and torques either as $\tau_1\dot{\theta}_1 + \cdots + \tau_6\dot{\theta}_6$ or, as in the previous section, in terms of velocities and force/torques in the tangent space of the group of rigid motions. Because we must get the same answer in either case we see that if the velocities $\dot{\theta}_1, \dot{\theta}_2, \ldots, \dot{\theta}_6$ are mapped into velocities in $\mathbf{E}(3)$ via the Jacobian df, then the force/torques in $\mathbf{E}(3)$ must be mapped into torques on the cotangent space of T^6 by the adjoint of df. *This is just the law of levers applied to our situation.* It has nothing to do with the fact that f is expressible as the product of exponentials.

What makes one kinematic chain better than another? There are, of course, many ways one could attempt to answer this question and, judging from the diversity of kinematic chains which one sees in the animal world, it seems unlikely that there is a universal answer. Nonetheless, since one can think of (df) as representing the "velocity gain" which corresponds to the map from joint coordinates to end effector coordinates, it follows that (when it exists) $((df)^*)^{-1}$ is the "torque gain" for the same map. Since we expect the robot to be able to move quickly and apply force at all points of the work space, one might begin with the idea that df and $((df)^*)^{-1}$ should be as close as possible to being isometries, i.e., the chain should come as close as possible to having the same "gain" everywhere in the workspace. We discuss a way to make this idea precise in the next paragraph.

The map corresponding to the kinematic chain sends a torus into a subset of $\mathbf{E}(3)$. In general this map will have singularities and the image will not be of the same topological type as the domain. Nonetheless we can attempt to formulate an integral criterion

$$\eta = \int \dots \int L(\theta_1, \theta_2, \dots, \theta_6)\, d\theta_1 \dots d\theta_6$$

whose minimization would result in making the velocity gain or force gain as uniform as possible. One way to do this which is quite natural begins by noticing that as far as the θ's are concerned, there is an obvious distance measure on the flat torus, given by

$$(ds)^2 = \sum_i (d\theta_i)^2$$

(If the servomotors which drive the revolute joints are of different sizes, this might be a weighted sum but we ignore this possible embellishment.) On the other hand, in $\mathbf{E}(3)$ there is a natural one parameter family of Riemannian structures as discussed above. If we pick one of these and let its matrix representation be denoted by G, then minimizing

$$\eta = \int \dots \int \operatorname{tr}[(df)^T G(df)]\, d\theta_1 \dots \theta_k$$

is a natural way to enforce a condition that df is never too large. However, notice that if, for example, we ask that $f : [0,1] \mapsto [0,1]$ include all of [0,1] in its image, then minimizing $\int_o^1 \left(\frac{df}{dx}\right)^2 dx$ results in $f(x) = x$. That is, df can't be small everywhere if f is an onto mapping. Thus η may be taken to be a measure of the uniformity of the gain of a kinematic chain. This is the functional which plays the critical role in the definition of harmonic maps [13].

Example: (See [11].) A planar mechanism with three revolute joints is shown in figure 3. The x, y coordinates of the end effector are given by

$$\begin{aligned} x &= l_1 \cos\theta_1 + l_2 \cos(\theta_1 + \theta_2) + l_3 \cos(\theta_1 + \theta_2 + \theta_3) \\ y &= l_1 \sin\theta_1 + l_2 \sin(\theta_1 + \theta_2) + l_3 \sin(\theta_1 + \theta_2 + \theta_3) \end{aligned}$$

This map takes T^3 into the disk of radius $l_1 + l_2 + l_3$ provided the lengths do not satisfy any inequality of the form

$$l_i + l_j < l_k$$

If we use the metric $(ds)^2 = (dx)^2 + (dy)^2$ in the disk, it may be shown that the choice of l_1, l_2, l_3 which minimizes η subject to $l_1 + l_2 + l_3 = l$ is $l_1 = 6l/11$ $l_2 = 3l/11$ $l_3 = 2l/11$. This can be compared with the lengths of the three segments of the human index finger.

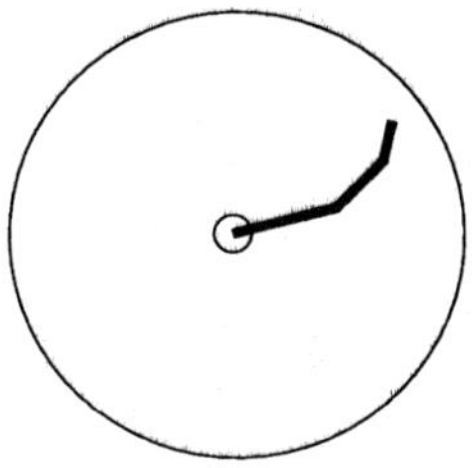

Figure 4. A three joint planar mechanism.

References

[1] Richard Paul, *Robot Manipulators: Mathematics, Programming and Control*, Cambridge, MA, MIT Press, 1981.

[2] R. W. Brockett, "Robotic Manipulations and the Product Exponentials Formulae," in *Lecture Notes in Control and Information Sciences, Proceedings of the International Symposium on Mathematical Theory of Networks and Systems*, Berlin, Springer-Verlag, 1984, pp. 120-127.

[3] K. H. Hunt, *Kinematic Geometry of Mechanism*, Oxford, England, Oxford University Press, 1978.

[4] J. M. Hervé, "Analyse structurelle des mécanismes par groupes de désplacements," *Mechanisms and Machine Theory*, Vol. 13, 1978, pp. 437-450.

[5] D. L. Pieper, *The Kinematics of Manipulation Under Computer Control*, Stanford Artificial Intelligence Laboratory, Stanford University, AIM 72, 1968.

[6] D. E. Whitney, "Quasi-static Assembly of Compliantly Supported Rigid Parts," *Journal of Dynamic Systems, Measurement and Control*, 1982, pp. 65-77.

[7] M. T. Mason, "Compliant Motion," in *Robot Motion Planning and Control*, (Brady *et al.*, eds.), Cambridge, MA, MIT Press, 1983.

[8] J. Loncaric, "Geometrical Analysis of Compliant Mechanisms in Robotics," Ph.D. Thesis, Division of Applied Sciences, Harvard University, 1985.

[9] T. Yoshikawa, "Manipulability of Robot Mechanism," *International Journal of Robotics Research*, Vol. 4-2, 1985.

[10] R. W. Brockett, and Josip Loncaric "The Geometry of Compliance Programming," *Theory and Application of Nonlinear Control Systems* (C. I. Byrnes and A. Lindquist, Eds.) Amsterdam: North Holland, 1986, pp. 35-42.

[11] Frank C. Park and R. W. Brockett, "Harmonic Maps and the Optimal Design of Mechanism," *Proceedings of the IEEE Conference on Decision and Control*, December 1989, pp.

[12] B. Paden and S. Sastry, "Optimal Kinematic Design of GR Manipulators," *Robotics Research*, Vol. 7, No. 2, 1988, pp. 43-61.

[13] J. Eells and J. H. Sampson, "Harmonic Mappings of Riemannian Manifolds," *American Journal of Mathematics*, Vol. 86, 1964, pp. 109-160.

[14] Willard Miller, Jr., *Symmetry Groups and their Applications*, Academic Press, New York, 1972.

[15] Robin J. Popplestone, Yanxi Liu, Rich Weiss, "A Group Theeoretic Approach to Assembly Planning," *AI Magazine*, Spring, 1990.

[16] V. Montanari and F. Rossi, "Fundamental Properties of Networks of Constraints: A New Formulation," in *Search in Artificial Intelligence* (L. Kanal and V. Kumar, Eds.) Springer-Verlag, New York, 1988.

[17] V. I. Arnold, *Mathematical Methods of Classical Mechanics*, Springer-Verlag, New York, 1978.

[18] R. Vijaykumar, K. J. Waldron, and M. J. tsai, "Geometric Optimization of Serial Chain Manipulator Structures for Working Volume and Dexterity," *International Journal of Robotics Research* Vol. 5-2, 1986.

[19] *Geometry and Robotics* (J. D. Boessennet and J. P. Laumond, Eds.), Springer-Verlag, Berlin, 1989.

[20] J. Michael McCarthy, *An Introduction to Theoretical Kinematics*, MIT Press, Cambridge, MA. 1990.

Proceedings of Symposia in Applied Mathematics
Volume **41**, 1990

Manipulator Kinematics

Madhusudan Raghavan

Abstract. We present the two fundamental problems in manipulator kinematics viz. the forward kinematics problem and the inverse kinematics problem. We show how these problems may be formulated as principal systems of multivariate polynomials. We describe Sylvester's dialytical elimination, an algebraic method for solving systems of multivariate polynomials. We then present some structure theorems on the equations describing series revolute-jointed manipulators. Finally we show how these theorems may be used along with the dialytical elimination method to solve the inverse kinematics problem for the six-revolute-jointed series manipulator of general geometry. The results presented in this manuscript are extensions of those described in RAGHAVAN and ROTH[1989].

1. Introduction. Manipulator kinematics is the branch of robotics which provides the analytical tools for describing a manipulator's geometry and motions. These tools are essential for robot motion planning and control. Formally kinematics is the branch of mechanics which treats the phenomenon of motion without regard to the cause of the motion. In kinematics there is no reference

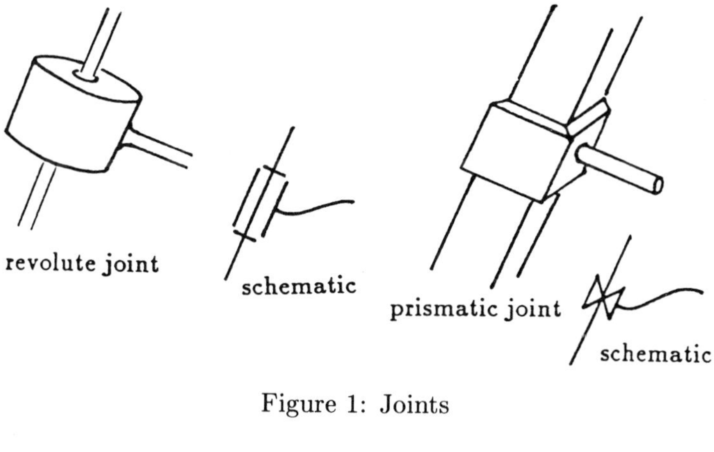

Figure 1: Joints

1980 *Mathematics Subject Classification* (1985 *Revision*). Primary 26B35, 26C10, 70B15.

to mass or force. The concern is only with relative positions and their changes. A manipulator is comprised of bodies or members which are connected to one another by joints. Most manipulators have revolute and prismatic joints. A revolute joint is similar to a hinge. It permits rotation of one body relative to another about an axis fixed in both bodies. No relative translation is permitted. A prismatic joint permits relative translation but no rotation. Figure 1 shows revolute and prismatic joints along with schematic or simplified diagrams. A manipulator usually has a special member which interacts with the environment and does useful work. This member is called the hand or end-effector. A manipulator consisting of a sequential chain of links and joints (like the human arm) is called a series manipulator. A manipulator in which two or more series chains of links and joints connect the hand to the ground is referred to as an in-parallel manipulator. A hybrid manipulator consists of combinations of series and in-parallel subsystems. Figure 2 shows schematic drawings of series, in-parallel,

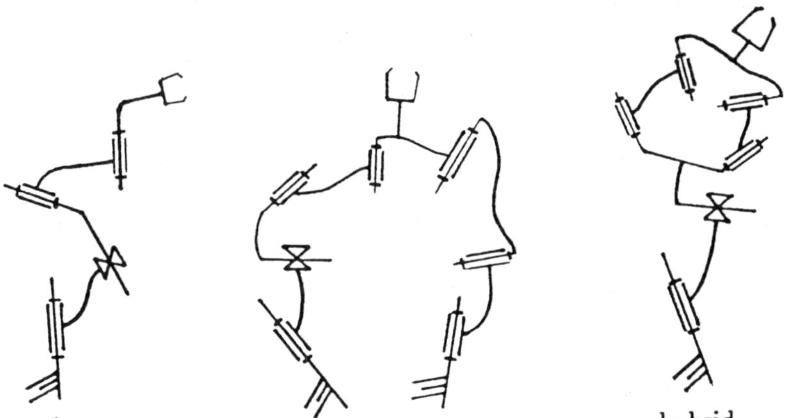

Figure 2: Examples

and hybrid manipulators. We restrict the discussion to series manipulators with rigid members and revolute joints. We present the two fundamental problems in manipulator kinematics, viz. the forward kinematics problem and the inverse kinematics problem. We show how the latter may be formulated as a system of n independent polynomials in n variables with the coefficients being real numbers. Such multivariate polynomial systems may be solved by algebraic elimination tools such as Gröbner Bases (BUCHBERGER[1985]), the multivariate resultant (CANNY[1988]) and Sylvester's dialytical method (SALMON[1885].) The most difficult problem in series manipulator inverse kinematics is that of the spatial six-revolute-jointed manipulator. We present a solution that uses dialytical elimination and exploits the structure of the polynomials describing the manipulator. To prepare the reader for this solution, we present a brief description of dialytical elimination. We then present some results on the structure of the polynomials describing a general spatial n-revolute-jointed series manipulator. Finally we show how these two sets of ideas may be combined into an elegant solution.

2. The forward and inverse kinematics problems. Although manipulators are in general spatial or three dimensional devices, we describe the forward and inverse kinematics problems with the help of a planar manipulator, i.e. a manipulator in which each link moves in a plane parallel to some reference plane fixed to the ground. This facilitates the presentation and there is no loss of generality since the fundamental concepts extend quite naturally to manipulators in which the links undergo three dimensional motion. Figure 3 is a schematic drawing of a planar manipulator with three revolute joints. The link with the

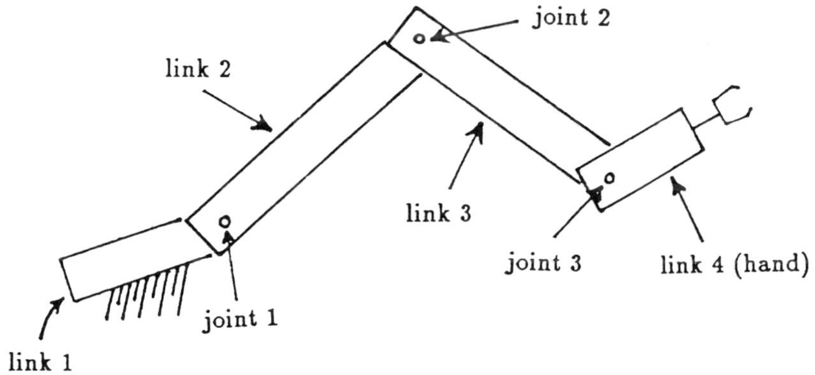

Figure 3: Planar Revolute-Jointed Manipulator

hatch marks is fixed to the ground and therefore remains stationary. Let us refer to it as link 1. Link 1 is connected to link 2 by means of a revolute joint (joint 1), the axis of which is normal to the plane of the paper. This joint permits link 2 to rotate relative to link 1. Similarly links 3 and 4 are connected by revolute joints 2 and 3 to links 2 and 3 respectively. The axes of joints 2 and 3 are also normal to the plane of the paper. Figure 4 is a side view of the planar manipulator showing the planes of motion of links 2, 3, and 4. These planes are parallel to each other. In order to develop a mathematical description of the manipulator we affix Cartesian coordinate systems to each link. These systems may be located and oriented arbitrarily on the links. We proceed with the arrangement

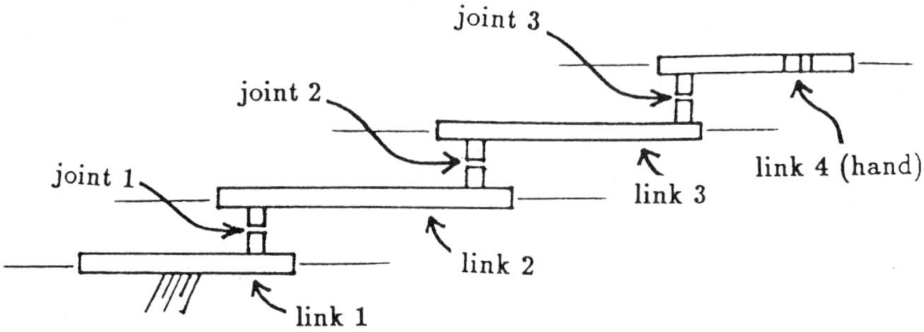

Figure 4: Side View of Planar Manipulator

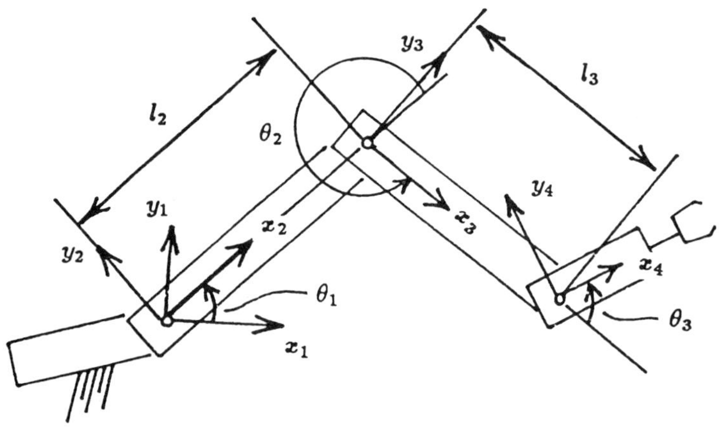

Figure 5: Planar Manipulator with Coordinate Systems

shown in Figure 5. Coordinate system x_i-y_i is located on link i, $i = 1, 2, 3, 4$. The origins of coordinate systems 1 and 2 are coincident. The position vector of the origin of coordinate system 3 in coordinate system 2 is $\begin{pmatrix} l_2 \\ 0 \end{pmatrix}$ and that of coordinate system 4 in system 3 is $\begin{pmatrix} l_3 \\ 0 \end{pmatrix}$. The forward and inverse kinematics problems are formulated as questions regarding the relative motions of these coordinate systems. The rotation of link $i+1$ relative to link i is measured by the angle θ_i. Let P be a point on link 4 with position vector $\mathbf{p}_4$ in coordinate system 4. Then $\mathbf{p}_3$ the position vector of P in coordinate system 3 is related to $\mathbf{p}_4$ by the following relation:

$$\mathbf{p}_3 = \begin{pmatrix} \cos\theta_3 & -\sin\theta_3 \\ \sin\theta_3 & \cos\theta_3 \end{pmatrix} \mathbf{p}_4 + \begin{pmatrix} l_3 \\ 0 \end{pmatrix}. \tag{1}$$

The reader may verify this by examining the vectors shown in Figure 6. (1) may be rewritten as

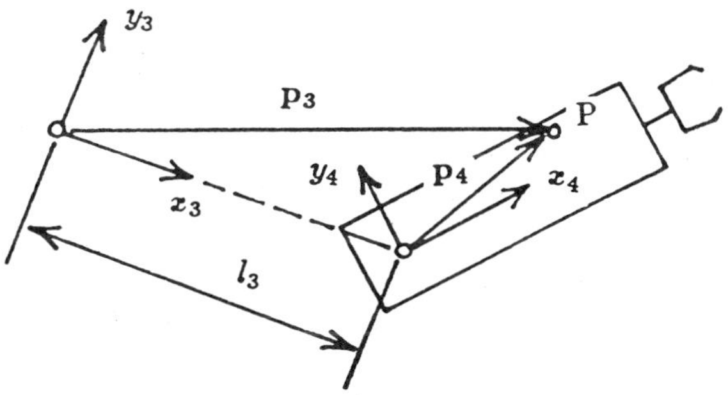

Figure 6: Position Vectors of P in Systems 3 and 4

$$\begin{pmatrix} \mathbf{p}_3 \\ 1 \end{pmatrix} = \begin{pmatrix} \cos\theta_3 & -\sin\theta_3 & l_3 \\ \sin\theta_3 & \cos\theta_3 & 0 \\ 0 & 0 & 1 \end{pmatrix} \begin{pmatrix} \mathbf{p}_4 \\ 1 \end{pmatrix}. \tag{2}$$

The matrix $\begin{pmatrix} \cos\theta_3 & -\sin\theta_3 & l_3 \\ \sin\theta_3 & \cos\theta_3 & 0 \\ 0 & 0 & 1 \end{pmatrix}$ permits a change of coordinates from system 4 to system 3. Let us refer to it as ${}^4_3\mathbf{T}$. By similar arguments ${}^3_2\mathbf{T}$, the coordinate transformation matrix from system 3 to system 2 is $\begin{pmatrix} \cos\theta_2 & -\sin\theta_2 & l_2 \\ \sin\theta_2 & \cos\theta_2 & 0 \\ 0 & 0 & 1 \end{pmatrix}$, and ${}^{2}_{1}\mathbf{T}$ is equal to $\begin{pmatrix} \cos\theta_1 & -\sin\theta_1 & 0 \\ \sin\theta_1 & \cos\theta_1 & 0 \\ 0 & 0 & 1 \end{pmatrix}$.

The matrices ${}^{i+1}_{i}\mathbf{T}, i = 1, 2, 3$, may be used to change coordinates from system 4 (the hand) to system 1 (the ground) by means of the following equation:

$$ {}^{2}_{1}\mathbf{T}\,{}^{3}_{2}\mathbf{T}\,{}^{4}_{3}\mathbf{T} = {}^{4}_{1}\mathbf{T}. \tag{3}$$

It governs the motion of the hand relative to the ground. The forward and inverse kinematics problems may now be defined with respect to (3). The forward problem is: given the lengths l_2, l_3, and the joints angles $\theta_1, \theta_2, \theta_3$, compute ${}^{4}_{1}\mathbf{T}$, the position and orientation of the hand relative to the ground. Clearly the forward kinematics problem is straightforward. It requires the evaluation of ${}^{i+1}_{i}\mathbf{T}, i = 1, 2, 3$, and the evaluation of the matrix product ${}^{2}_{1}\mathbf{T}\,{}^{3}_{2}\mathbf{T}\,{}^{4}_{3}\mathbf{T}$. On the other hand the inverse kinematics problem is fairly complicated. The problem statement is: given l_2, l_3, and ${}^{4}_{1}\mathbf{T}$, the hand position and orientation relative to the ground, find the corresponding values of the joint angles $\theta_1, \theta_2, \theta_3$. Equating the entries on the left and right-hand sides of rows 1 and 2 in (3) yields six equations in the three variables $\theta_1, \theta_2, \theta_3$. Only three of these equations are independent. This is because the (1,1), (1,2), (2,1), (2,2) submatrix on either side of (3) is an element of $\mathcal{SO}(2)$, the group of orthogonal 2×2 matrices with determinant equal to $+1$.[1] Therefore only one of the four entries in this submatrix is independent. Equating the left and right-hand sides of this submatrix in (3) yields only one independent equation instead of four. Any one of the four equations may be treated as the independent one. Let us arbitrarily pick the (2,2) equation. This equation along with the (1,3) and (2,3) equations give us a system of three independent equations in the three variables $\theta_1, \theta_2, \theta_3$. These three equations are

$$ l_2 \cos\theta_1 + l_3 \cos(\theta_1 + \theta_2) = {}^{4}_{1}\mathbf{t}_{13}, \tag{4}$$

$$ l_2 \sin\theta_1 + l_3 \sin(\theta_1 + \theta_2) = {}^{4}_{1}\mathbf{t}_{23}, \tag{5}$$

$$ \cos(\theta_1 + \theta_2 + \theta_3) = {}^{4}_{1}\mathbf{t}_{22}, \tag{6}$$

where ${}^{4}_{1}\mathbf{t}_{ij}$ is the $(i,j)^{th}$ entry of ${}^{4}_{1}\mathbf{T}$. The inverse kinematics problem requires that we solve these three equations for θ_1, θ_2, and θ_3. (4), (5), and (6) may be rewritten as polynomials by making the substitutions $\sin\theta_i = x_i$, $\cos\theta_i = y_i$, with the additional relation $x_i^2 + y_i^2 = 1$. The resulting system of six polynomials in the six variables $x_1, y_1, x_2, y_2, x_3, y_3$, is

$$ l_2 y_1 + l_3(y_1 y_2 - x_1 x_2) = {}^{4}_{1}\mathbf{t}_{13}, \tag{7}$$

$$ l_2 x_1 + l_3(y_1 x_2 + x_1 y_2) = {}^{4}_{1}\mathbf{t}_{23}, \tag{8}$$

$$ -x_1(y_2 x_3 + x_2 y_3) - y_1(x_2 x_3 - y_2 y_3) = {}^{4}_{1}\mathbf{t}_{22}, \tag{9}$$

[1]The reader should verify that the (1,1), (1,2), (2,1), (2,2) submatrix in ${}^{i+1}_{i}\mathbf{T}, i = 1, 2, 3$, is an element of $\mathcal{SO}(2)$. The reader should also verify that the matrix obtained by multiplying any two of these transformation matrices also exhibits the same property.

$$x_1^2 + y_1^2 = 1, \tag{10}$$

$$x_2^2 + y_2^2 = 1, \tag{11}$$

$$x_3^2 + y_3^2 = 1. \tag{12}$$

An alternate substitution is $\sin\theta_i = \frac{2x_i}{1+x_i^2}$, $\cos\theta_i = \frac{1-x_i^2}{1+x_i^2}$, where $x_i = \tan(\frac{\theta_i}{2})$. This would give a system of three rational functions in the three variables x_1, x_2, x_3. After clearing denominators this system could be treated as a set of three polynomials in x_1, x_2, x_3. Systems of nonlinear multivariate polynomials such as these typically have several common zeros. For our inverse kinematics problem the physical significance of several $(\theta_1, \theta_2, \theta_3)$ tuples satisfying (4), (5), and (6) is that the manipulator can place its hand in the position and orientation specified by ${}^4_1\mathbf{T}$, in more ways than one. Figure 7 shows two ways (or configurations) in which a planar three-revolute-jointed manipulator places its hand in a specified position and orientation. A knowledge of all solutions to an inverse kinematics problem is generally desirable for motion planning because the presence of obstacles could rule out the use of some of the solutions (see Figure 8.) In summary, the important issues in any manipulator inverse kinematics

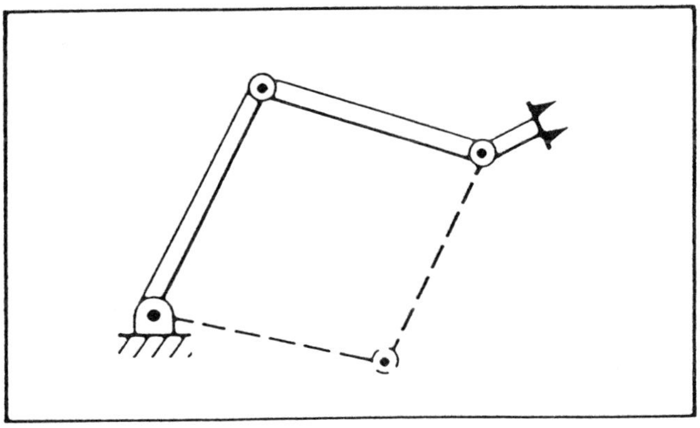

Figure 7: Two Inverse Kinematic Solutions

problem are: (i) how many different configurations of the manipulator permit its hand to occupy a desired position and orientation ? (ii) how does one compute these configurations ? The procedure to be followed in any inverse kinematics problem is the same as the one used for our planar three-revolute-jointed manipulator. The first step is to affix Cartesian coordinate systems on each link of the manipulator. The next step requires writing down the coordinate transformation

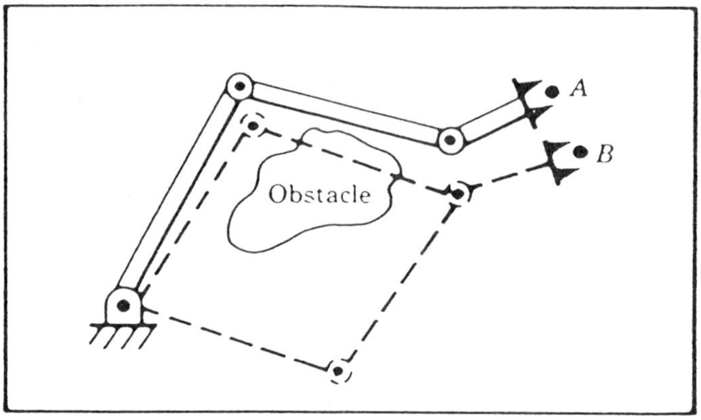

One of the two possible solutions to reach point B from point A causes a collision.

Figure 8: Obstacle Avoidance

matrices relating adjacent Cartesian coordinate systems. These matrices may then be used in a matrix equation of the form

$$ {}^2_1\mathbf{T}\,{}^3_2\mathbf{T} \ldots {}^{hand}_{n}\mathbf{T} = {}^{hand}_{1}\mathbf{T}, \tag{13} $$

relating the coordinate sytem on the hand to the coordinate system on the ground. The numerical value of ${}^{hand}_{1}\mathbf{T}$, the matrix describing the *desired* position and orientation of the hand, would be specified in an inverse kinematics problem. Numerical values of the invariant parameters of the manipulator such as the lengths of its links, would also be specified. The above matrix equation must then be rewritten as a system of multivariate polynomials to be solved for the joint variables. Algebraic methods for solving systems of multivariate polynomials proceed by eliminating variables and reducing the problem to the solution of a univariate polynomial. The elimination of variables may be effected by Gröbner Bases, resultants and by dialytical elimination. Elimination by Gröbner Bases and resultants generally requires a large number of computations for even fairly small problems. Dialytical elimination on the other hand is quite efficient for polynomial systems exhibiting sparsity and structure, though in the worst case it could be as tedious as elimination by resultants. We now present a brief description of the dialytical elimination procedure.

3. Dialytical elimination. We describe this elimination procedure with the help of the following system of three non-homogeneous polynomials f_1, f_2, f_3, in the three variables x_1, x_2, x_3:

$$\begin{pmatrix} f_1 \\ f_2 \\ f_3 \end{pmatrix} = \begin{pmatrix} a_1 & b_1 & c_1 & d_1 & e_1 & g_1 \\ a_2 & b_2 & c_2 & d_2 & e_2 & g_2 \\ a_3 & b_3 & c_3 & d_3 & e_3 & g_3 \end{pmatrix} \begin{pmatrix} x_1^2 \\ x_1x_2 \\ x_1x_3 \\ x_2^2 \\ x_3 \\ 1 \end{pmatrix} = \begin{pmatrix} 0 \\ 0 \\ 0 \end{pmatrix}. \tag{14}$$

The coefficients $a_i, \ldots, g_i, i = 1, 2, 3$, may be arbitrary complex numbers. These polynomials are to be solved for their common zeros and we do so by eliminating x_2 and x_3 so as to obtain a polynomial in x_1 which vanishes at the common zeros of f_1, f_2, f_3.
We rewrite (14) as a system of three non-homogeneous polynomials in x_2, x_3, over the ring $\mathbf{C}[x_1]$ as follows:

$$\begin{pmatrix} f_1 \\ f_2 \\ f_3 \end{pmatrix} = \begin{pmatrix} d_1 & b_1x_1 & e_1 + c_1x_1 & a_1x_1^2 + g_1 \\ d_2 & b_2x_1 & e_2 + c_2x_1 & a_2x_1^2 + g_2 \\ d_3 & b_3x_1 & e_3 + c_3x_1 & a_3x_1^2 + g_3 \end{pmatrix} \begin{pmatrix} x_2^2 \\ x_2 \\ x_3 \\ 1 \end{pmatrix} = \begin{pmatrix} 0 \\ 0 \\ 0 \end{pmatrix}. \tag{15}$$

Multiply f_1, f_2, f_3, by powers of x_2 and x_3 so as to generate new elements in the ideal of f_1, f_2, f_3, which are linearly independent of f_1, f_2, f_3.[2] Generate as many elements as are required to form a square coefficient array. For example, multiplying f_1, f_2, f_3, by x_2 yields the following three elements in the ideal of f_1, f_2, f_3:

$$\begin{pmatrix} x_2f_1 \\ x_2f_2 \\ x_2f_3 \end{pmatrix} = \begin{pmatrix} d_1 & b_1x_1 & e_1 + c_1x_1 & a_1x_1^2 + g_1 \\ d_2 & b_2x_1 & e_2 + c_2x_1 & a_2x_1^2 + g_2 \\ d_3 & b_3x_1 & e_3 + c_3x_1 & a_3x_1^2 + g_3 \end{pmatrix} \begin{pmatrix} x_2^3 \\ x_2^2 \\ x_2x_3 \\ x_2 \end{pmatrix} = \begin{pmatrix} 0 \\ 0 \\ 0 \end{pmatrix}. \tag{16}$$

These elements are linearly independent of f_1, f_2, f_3, because clearly they cannot be expressed in the form $af_1 + bf_2 + cf_3$, where a, b, c, are elements of $\mathbf{C}[x_1]$. (15) and (16) taken together may be written as:

$$\begin{pmatrix} f_1 \\ f_2 \\ f_3 \\ x_2f_1 \\ x_2f_2 \\ x_2f_3 \end{pmatrix} =$$

[2]By the ideal of f_1, f_2, f_3, we mean the set of all polynomials of the form $q_1f_1 + q_2f_2 + q_3f_3$, where q_1, q_2, q_3, are arbitrary elements of the set of polynomials in x_2, x_3, over the ring $\mathbf{C}[x_1]$.

$$\begin{pmatrix} 0 & d_1 & 0 & b_1x_1 & e_1+c_1x_1 & a_1x_1^2+g_1 \\ 0 & d_2 & 0 & b_2x_1 & e_2+c_2x_1 & a_2x_1^2+g_2 \\ 0 & d_3 & 0 & b_3x_1 & e_3+c_3x_1 & a_3x_1^2+g_3 \\ d_1 & b_1x_1 & e_1+c_1x_1 & a_1x_1^2+g_1 & 0 & 0 \\ d_2 & b_2x_1 & e_2+c_2x_1 & a_2x_1^2+g_2 & 0 & 0 \\ d_3 & b_3x_1 & e_3+c_3x_1 & a_3x_1^2+g_3 & 0 & 0 \end{pmatrix} \begin{pmatrix} x_2^3 \\ x_2^2 \\ x_2x_3 \\ x_2 \\ x_3 \\ 1 \end{pmatrix} = \begin{pmatrix} 0 \\ 0 \\ 0 \\ 0 \\ 0 \\ 0 \end{pmatrix}. \tag{17}$$

The coefficient matrix is square and therefore we need not generate any more elements in the ideal of f_1, f_2, f_3. The determinant of the coefficient matrix in (17) must vanish at each root of f_1, f_2, f_3. If it does not vanish, the coefficient matrix is invertible at that value of x_1. Multiplying both sides of (17) by the inverse of the coefficient matrix yields:

$$\begin{pmatrix} x_2^3 \\ x_2^2 \\ x_2x_3 \\ x_2 \\ x_3 \\ 1 \end{pmatrix} = \begin{pmatrix} 0 \\ 0 \\ 0 \\ 0 \\ 0 \\ 0 \end{pmatrix}. \tag{18}$$

The last row of (18) is a contradiction. Therefore the determinant of the coefficient matrix in (17) is the desired univariate polynomial in x_1. After computing the roots of this polynomial we may compute the values of x_2 and x_3 corresponding to each value of x_1 by substituting for x_1 in the coefficient matrix of (17) to obtain

$$A \begin{pmatrix} x_2^3 \\ x_2^2 \\ x_2x_3 \\ x_2 \\ x_3 \\ 1 \end{pmatrix} = \begin{pmatrix} 0 \\ 0 \\ 0 \\ 0 \\ 0 \\ 0 \end{pmatrix}, \tag{19}$$

where A is a 6×6 array of complex numbers. We may pick any five independent equations in (19), treat them as a linear system in five variables, and solve for $x_2^3, x_2^2, x_2x_3, x_2, x_3$. The drawback of the dialytical elimination method is that the univariate polynomial in x_1 may be of degree higher than the number of common zeros of f_1, f_2, f_3. This procedure may therefore yield extraneous roots along with the true ones. This may be remedied by substituting each candidate root back into f_1, f_2, f_3 to see whether they vanish. Additional details of this elimination procedure may be found in SALMON[1885], pages 79-90. James Joseph Sylvester introduced the name "dialytical" because the process dissolves the relations which connect the different combinations of powers of the variables and treats them as simple independent quantities.

4. Properties of the n-revolute-jointed manipulator. The position and orientation of a body in three dimensional space ($\Re^3$) is determined by six parameters; three for position and three for orientation. A manipulator capable of positioning and orienting a body (its hand) arbitrarily in three space must therefore possess at least six degrees of freedom. The number of degrees of freedom of a manipulator is the number of independent parameters which would have to be specified in order to locate all parts of the manipulator. In a series manipulator with revolute and prismatic joints each joint contributes one independent parameter viz. the joint rotation angle or the joint displacement, depending on whether the joint is revolute or prismatic. Therefore a six-degrees-of-freedom series manipulator with revolute and prismatic joints must have a total of six joints. The inverse kinematics of all six-degrees-of-freedom series manipulators with at least one prismatic joint has been solved over the past twenty years. However the inverse kinematics of the series manipulator with six revolute joints remained unsolved until 1988 (see LEE and LIANG[1988a and 1988b].) The reason for this is the excessive complexity of the variable elimination process for this manipulator. This problem was appropriately named "the Mount Everest of series manipulator inverse kinematics," by FREUDENSTEIN[1973]. Recently we derived a new solution to this problem. In the present section we develop the tools for our solution. We then show how they may be combined with the dialytical elimination procedure to solve this problem.

4.1 Notation for spatial manipulators. There is a natural extension of the ideas in Section 2 to spatial or three dimensional manipulators. For such manipulators the Cartesian coordinate system attached to each link is three dimensional. The coordinate transformation matrices ${}^{i+1}_{i}\mathbf{T}$ are of size 4×4. Though the coordinate systems may be located and oriented arbitrarily on each link, the research community has agreed upon a set of rules with a view to standardizing notation. These rules are explained in detail in TSAI and MORGAN[1985] and the essential features are as follows:

1. The links are numbered in ascending order outwards with the fixed link being number 1. Joint i connects links i and $i+1$.
2. Coordinate system i is located on link i with its z axis (z_i) along the axis of joint i. The x_i axis is along the common normal to the axes of joints i and $i-1$. The y_i axis is normal to the x_i and z_i axes.
3. The first coordinate system is fixed to the ground with the x_1 axis directed arbitrarily in the plane normal to the z_1 axis.
4. The coordinate sytem on the hand may be located arbitrarily. The z axis direction may also be selected arbitrarily. The x axis is defined by the common normal to the z axis on the last link and the z axis on the preceding coordinate system. The y axis is normal to the x and z axes.

For such an arrangement the coordinate transformation matrix ${}^{i+1}_{i}\mathbf{T}$ is

$$\begin{pmatrix} \cos\theta_i & -\sin\theta_i\cos\alpha_i & \sin\theta_i\sin\alpha_i & a_i\cos\theta_i \\ \sin\theta_i & \cos\theta_i\cos\alpha_i & -\cos\theta_i\sin\alpha_i & a_i\sin\theta_i \\ 0 & \sin\alpha_i & \cos\alpha_i & d_i \\ 0 & 0 & 0 & 1 \end{pmatrix},$$

where θ, α, a, and d are shown in Figure 9. A formal derivation of ${}^{i+1}_{i}\mathbf{T}$ may be

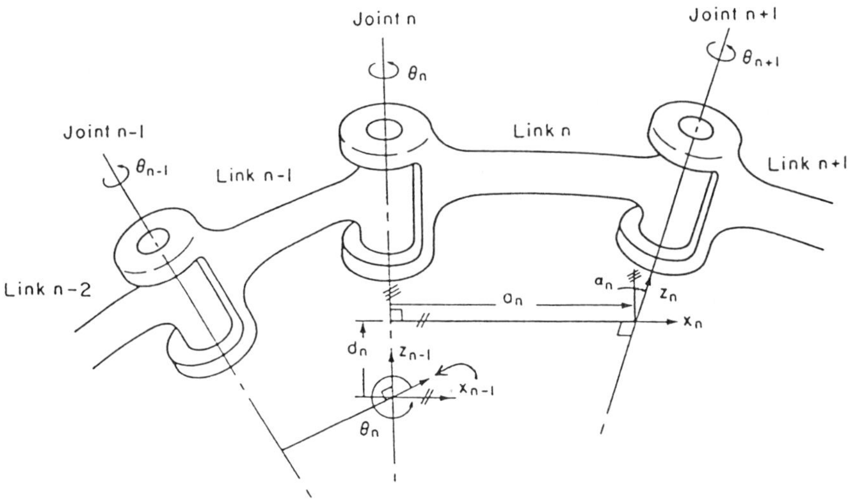

Figure 9: Kinematic Parameters

found in PAUL[1981], Chapter 2, page 53. For a revolute-jointed manipulator, θ_i is the angle of rotation of link $i+1$ relative to link i, about joint i. Unlike θ_i, the parameters α_i, a_i and d_i are constants. The transformation ${}^{n+1}_{1}\mathbf{T}$ for a n-revolute-jointed spatial series manipulator is related to the individual coordinate transformation matrices ${}^{i+1}_{i}\mathbf{T}$, $i = 1, \ldots, n$, by the following equation:

$$ {}^{2}_{1}\mathbf{T}\,{}^{3}_{2}\mathbf{T} \ldots {}^{n+1}_{n}\mathbf{T} = {}^{n+1}_{1}\mathbf{T}. \tag{20}$$

Let ${}^{i}_{j}\mathbf{T}$, (i, j, integers) be written as $\begin{pmatrix} {}^{i}_{j}\mathbf{R} & {}^{i}_{j}\mathbf{p} \\ 0 & 1 \end{pmatrix}$ where ${}^{i}_{j}\mathbf{R}$ is a 3×3 matrix comprised of the first three rows and columns of ${}^{i}_{j}\mathbf{T}$, and $\begin{pmatrix} {}^{i}_{j}\mathbf{p} \\ 1 \end{pmatrix}$ is column 4 of ${}^{i}_{j}\mathbf{T}$. The reader should verify that the relation ${}^{i}_{j}\mathbf{T} = {}^{j+1}_{j}\mathbf{T}\,{}^{j+2}_{j+1}\mathbf{T} \ldots {}^{i}_{i-1}\mathbf{T}$, implies that ${}^{i}_{j}\mathbf{R} = {}^{j+1}_{j}\mathbf{R}\,{}^{j+2}_{j+1}\mathbf{R} \ldots {}^{i}_{i-1}\mathbf{R}$. By definition, ${}^{j+1}_{j}\mathbf{R}$ is an element of $\mathcal{SO}(3)$ [3] for

[3] $\mathcal{SO}(3)$ is the group of orthogonal 3×3 matrices with determinant equal to $+1$.

all j. Let ${}^i_j\mathbf{R}$ be written as $\begin{pmatrix} {}^i_j\mathbf{l} & {}^i_j\mathbf{m} & {}^i_j\mathbf{n} \end{pmatrix}$ where ${}^i_j\mathbf{l}$, ${}^i_j\mathbf{m}$, ${}^i_j\mathbf{n}$, are respectively columns 1, 2, and 3 of ${}^i_j\mathbf{R}$. (20) may be rewritten as:

$$ {}^n_1\mathbf{T}{}^{n+1}_n\mathbf{T} = {}^{n+1}_1\mathbf{T}. \tag{21} $$

Making the following substitutions in (21)

$$ {}^n_1\mathbf{T} \leftarrow \begin{pmatrix} {}^n_1\mathbf{R} & {}^n_1\mathbf{p} \\ 0 & 1 \end{pmatrix}, $$

$$ {}^{n+1}_n\mathbf{T} \leftarrow \begin{pmatrix} {}^{n+1}_n\mathbf{l} & {}^{n+1}_n\mathbf{m} & {}^{n+1}_n\mathbf{n} & {}^{n+1}_n\mathbf{p} \\ 0 & 0 & 0 & 1 \end{pmatrix}, $$

$$ {}^{n+1}_1\mathbf{T} \leftarrow \begin{pmatrix} {}^{n+1}_1\mathbf{l} & {}^{n+1}_1\mathbf{m} & {}^{n+1}_1\mathbf{n} & {}^{n+1}_1\mathbf{p} \\ 0 & 0 & 0 & 1 \end{pmatrix}, $$

we get

$$ \begin{pmatrix} {}^n_1\mathbf{R} & {}^n_1\mathbf{p} \\ 0 & 1 \end{pmatrix} \begin{pmatrix} {}^{n+1}_n\mathbf{l} & {}^{n+1}_n\mathbf{m} & {}^{n+1}_n\mathbf{n} & {}^{n+1}_n\mathbf{p} \\ 0 & 0 & 0 & 1 \end{pmatrix} = \begin{pmatrix} {}^{n+1}_1\mathbf{l} & {}^{n+1}_1\mathbf{m} & {}^{n+1}_1\mathbf{n} & {}^{n+1}_1\mathbf{p} \\ 0 & 0 & 0 & 1 \end{pmatrix}. \tag{22} $$

Equating columns 1, 2, 3, and 4 on either side of (22) we get

$$ {}^n_1\mathbf{R}{}^{n+1}_n\mathbf{l} = {}^{n+1}_1\mathbf{l}, \tag{23} $$

$$ {}^n_1\mathbf{R}{}^{n+1}_n\mathbf{m} = {}^{n+1}_1\mathbf{m}, \tag{24} $$

$$ {}^n_1\mathbf{R}{}^{n+1}_n\mathbf{n} = {}^{n+1}_1\mathbf{n}, \tag{25} $$

$$ {}^n_1\mathbf{R}{}^{n+1}_n\mathbf{p} + {}^n_1\mathbf{p} = {}^{n+1}_1\mathbf{p}. \tag{26} $$

We now present some results on the structure of (20)-(26). The formal proofs supporting these results are available in Appendix 1.

4.2 Structure theorems.

LEMMA. *The equations obtained by equating entries from the left and right-hand sides of (20) are of degree 1 or less in* $\sin\theta_i, \cos\theta_i, i = 1, 2, \ldots, n$.

This means that for any given integer i, the highest power of $\sin\theta_i, \cos\theta_i$, appearing in these equations is less than or equal to 1.

PROPERTY 1. $({}^{n+1}_1\mathbf{p}) \cdot ({}^{n+1}_1\mathbf{p})$ *is of degree 1 or less in* $\sin\theta_i, \cos\theta_i, i = 1, 2, \ldots, n$.

PROPERTY 2. *Each of the functions* $({}^{n+1}_1\mathbf{p})\cdot({}^{n+1}_1\mathbf{l})$, $({}^{n+1}_1\mathbf{p})\cdot({}^{n+1}_1\mathbf{m})$, $({}^{n+1}_1\mathbf{p})\cdot({}^{n+1}_1\mathbf{n})$, *is of degree 1 or less in* $\sin\theta_i, \cos\theta_i, i = 1, 2, \ldots, n$.

PROPERTY 3. *The entries of the vectors* ${}_{1}^{n+1}\mathbf{p}\times{}_{1}^{n+1}\mathbf{l}$, ${}_{1}^{n+1}\mathbf{p}\times{}_{1}^{n+1}\mathbf{m}$, ${}_{1}^{n+1}\mathbf{p}\times{}_{1}^{n+1}\mathbf{n}$ *are of degree 1 or less in* $\sin\theta_i, \cos\theta_i, i = 1, 2, \ldots, n$.

PROPERTY 4. *The functions* $({}_{1}^{n+1}\mathbf{p}\cdot{}_{1}^{n+1}\mathbf{p}){}_{1}^{n+1}\mathbf{l} - (2{}_{1}^{n+1}\mathbf{p}\cdot{}_{1}^{n+1}\mathbf{l}){}_{1}^{n+1}\mathbf{p}$, $({}_{1}^{n+1}\mathbf{p}\cdot{}_{1}^{n+1}\mathbf{p}){}_{1}^{n+1}\mathbf{m} - (2{}_{1}^{n+1}\mathbf{p}\cdot{}_{1}^{n+1}\mathbf{m}){}_{1}^{n+1}\mathbf{p}$, *and* $({}_{1}^{n+1}\mathbf{p}\cdot{}_{1}^{n+1}\mathbf{p}){}_{1}^{n+1}\mathbf{n} - (2{}_{1}^{n+1}\mathbf{p}\cdot{}_{1}^{n+1}\mathbf{n}){}_{1}^{n+1}\mathbf{p}$ *are of degree 1 or less in* $\sin\theta_i, \cos\theta_i, i = 1, 2, \ldots, n$.

5. Six-revolute-jointed manipulator inverse kinematics. We now show how the properties presented in the previous section may be used along with the dialytical elimination procedure to solve the inverse kinematics problem for the spatial six-revolute-jointed manipulator (6R) of general geometry. (20), with n set equal to 6 is to be solved for the six variables $\theta_1, \ldots, \theta_6$. Numerical values for $a_i, \alpha_i, d_i, i = 1, \ldots, 6$, and ${}_{1}^{7}\mathbf{T}$ are provided and are arbitrarily real numbers. Let ${}_{1}^{7}\mathbf{T}$ be equal to $\begin{pmatrix} l_x & m_x & n_x & p_x \\ l_y & m_y & n_y & p_y \\ l_z & m_z & n_z & p_z \\ 0 & 0 & 0 & 1 \end{pmatrix}$. (20) may be rewritten as

$$ {}_{3}^{4}\mathbf{T}{}_{4}^{5}\mathbf{T}{}_{5}^{6}\mathbf{T} = {}_{2}^{3}\mathbf{T}^{-1}{}_{1}^{2}\mathbf{T}^{-1}{}_{1}^{hand}\mathbf{T}{}_{6}^{7}\mathbf{T}^{-1}. \tag{27} $$

We do this so as to move θ_1, θ_2, and θ_6 to the right-hand side. This lowers the degrees of the equations and also reduces their complexity.

$$ \begin{pmatrix} (\theta_3,\theta_4,\theta_5) & (\theta_3,\theta_4,\theta_5) & (\theta_3,\theta_4,\theta_5) & (\theta_3,\theta_4,\theta_5) \\ (\theta_3,\theta_4,\theta_5) & (\theta_3,\theta_4,\theta_5) & (\theta_3,\theta_4,\theta_5) & (\theta_3,\theta_4,\theta_5) \\ (\theta_4,\theta_5) & (\theta_4,\theta_5) & (\theta_4,\theta_5) & (\theta_4,\theta_5) \\ 0 & 0 & 0 & 1 \end{pmatrix} = $$

$$ \begin{pmatrix} (\theta_1,\theta_2,\theta_6) & (\theta_1,\theta_2,\theta_6) & (\theta_1,\theta_2) & (\theta_1,\theta_2) \\ (\theta_1,\theta_2,\theta_6) & (\theta_1,\theta_2,\theta_6) & (\theta_1,\theta_2) & (\theta_1,\theta_2) \\ (\theta_1,\theta_2,\theta_6) & (\theta_1,\theta_2,\theta_6) & (\theta_1,\theta_2) & (\theta_1,\theta_2) \\ 0 & 0 & 0 & 1 \end{pmatrix}. \tag{28} $$

(28) shows the variables appearing in the individual entries in (27). The six scalar equations obtained from columns 4 and 3 of (27) are devoid of θ_6. These equations though linearly independent are governed by the constraint that the magnitude of the column 3 vector is unity. We work with these six equations with the goal of eliminating four of the five variables so as to obtain a univariate polynomial which will vanish at their common zeros. The six equations are (note: $\sin\theta_i$ is denoted by s_i, $\cos\theta_i$ by c_i, $\sin\alpha_i$ by μ_i, and $\cos\alpha_i$ by λ_i.):

$$ c_3 f_1 + s_3 f_2 = c_2 h_1 + s_2 h_2 - a_2, \tag{29} $$

$$ s_3 f_1 - c_3 f_2 = -\lambda_2(s_2 h_1 - c_2 h_2) + \mu_2(h_3 - d_2), \tag{30} $$

$$ f_3 = \mu_2(s_2 h_1 - c_2 h_2) + \lambda_2(h_3 - d_2), \tag{31} $$

$$ c_3 r_1 + s_3 r_2 = c_2 b_1 + s_2 b_2, \tag{32} $$

$$s_3 r_1 - c_3 r_2 = -\lambda_2(s_2 b_1 - c_2 b_2) + \mu_2 b_3, \tag{33}$$

$$r_3 = \mu_2(s_2 b_1 - c_2 b_2) + \lambda_2 b_3, \tag{34}$$

where

$$\begin{aligned} f_1 &= c_4 g_1 + s_4 g_2 + a_3, \\ f_2 &= -\lambda_3(s_4 g_1 - c_4 g_2) + \mu_3 g_3, \\ f_3 &= \mu_3(s_4 g_1 - c_4 g_2) + \lambda_3 g_3 + d_3, \end{aligned}$$

$$\begin{aligned} r_1 &= c_4 m_1 + s_4 m_2, \\ r_2 &= -\lambda_3(s_4 m_1 - c_4 m_2) + \mu_3 m_3, \\ r_3 &= \mu_3(s_4 m_1 - c_4 m_2) + \lambda_3 m_3, \end{aligned}$$

$$\begin{aligned} g_1 &= c_5 a_5 + a_4, \\ g_2 &= -s_5 \lambda_4 a_5 + \mu_4 d_5, \\ g_3 &= s_5 \mu_4 a_5 + \lambda_4 d_5 + d_4, \end{aligned}$$

$$\begin{aligned} m_1 &= s_5 \mu_5, \\ m_2 &= c_5 \lambda_4 \mu_5 + \mu_4 \lambda_5, \\ m_3 &= -c_5 \mu_4 \mu_5 + \lambda_4 \lambda_5, \end{aligned}$$

$$\begin{aligned} h_1 &= c_1 p + s_1 q - a_1, \\ h_2 &= -\lambda_1(s_1 p - c_1 q) + \mu_1(r - d_1), \\ h_3 &= \mu_1(s_1 p - c_1 q) + \lambda_1(r - d_1), \end{aligned}$$

$$\begin{aligned} b_1 &= c_1 u + s_1 v, \\ b_2 &= -\lambda_1(s_1 u - c_1 v) + \mu_1 w, \\ b_3 &= \mu_1(s_1 u - c_1 v) + \lambda_1 w, \end{aligned}$$

$$\begin{aligned} p &= -l_x a_6 - (m_x \mu_6 + n_x \lambda_6) d_6 + \rho_x, \\ q &= -l_y a_6 - (m_y \mu_6 + n_y \lambda_6) d_6 + \rho_y, \\ r &= -l_z a_6 - (m_z \mu_6 + n_z \lambda_6) d_6 + \rho_z, \end{aligned}$$

$$\begin{aligned} u &= m_x \mu_6 + n_x \lambda_6, \\ v &= m_y \mu_6 + n_y \lambda_6, \\ w &= m_z \mu_6 + n_z \lambda_6. \end{aligned}$$

(29)-(34) may be written in matrix form as follows:

$$\begin{pmatrix} c_3 & s_3 & 0 \\ s_3 & -c_3 & 0 \\ 0 & 0 & 1 \end{pmatrix} \begin{pmatrix} f_1 \\ f_2 \\ f_3 \end{pmatrix} = \begin{pmatrix} 1 & 0 & 0 \\ 0 & -\lambda_2 & \mu_2 \\ 0 & \mu_2 & \lambda_2 \end{pmatrix} \left\{ \begin{pmatrix} c_2 & s_2 & 0 \\ s_2 & -c_2 & 0 \\ 0 & 0 & 1 \end{pmatrix} \begin{pmatrix} h_1 \\ h_2 \\ h_3 \end{pmatrix} + \begin{pmatrix} -a_2 \\ 0 \\ -d_2 \end{pmatrix} \right\}, \tag{35}$$

$$\begin{pmatrix} c_3 & s_3 & 0 \\ s_3 & -c_3 & 0 \\ 0 & 0 & 1 \end{pmatrix} \begin{pmatrix} r_1 \\ r_2 \\ r_3 \end{pmatrix} = \begin{pmatrix} 1 & 0 & 0 \\ 0 & -\lambda_2 & \mu_2 \\ 0 & \mu_2 & \lambda_2 \end{pmatrix} \begin{pmatrix} c_2 & s_2 & 0 \\ s_2 & -c_2 & 0 \\ 0 & 0 & 1 \end{pmatrix} \begin{pmatrix} b_1 \\ b_2 \\ b_3 \end{pmatrix}. \tag{36}$$

Premultiplying (35) by $\begin{pmatrix} 1 & 0 & 0 \\ 0 & -\lambda_2 & \mu_2 \\ 0 & \mu_2 & \lambda_2 \end{pmatrix}$ and adding $\begin{pmatrix} a_2 \\ 0 \\ d_2 \end{pmatrix}$ we get

$$\begin{pmatrix} c_2 & s_2 & 0 \\ s_2 & -c_2 & 0 \\ 0 & 0 & 1 \end{pmatrix} \tilde{\mathbf{h}} = \begin{pmatrix} 1 & 0 & 0 \\ 0 & -\lambda_2 & \mu_2 \\ 0 & \mu_2 & \lambda_2 \end{pmatrix} \begin{pmatrix} c_3 & s_3 & 0 \\ s_3 & -c_3 & 0 \\ 0 & 0 & 1 \end{pmatrix} \tilde{\mathbf{f}} + \begin{pmatrix} a_2 \\ 0 \\ d_2 \end{pmatrix}, \tag{37}$$

where $\tilde{\mathbf{h}} = \begin{pmatrix} h_1 \\ h_2 \\ h_3 \end{pmatrix}$ and $\tilde{\mathbf{f}} = \begin{pmatrix} f_1 \\ f_2 \\ f_3 \end{pmatrix}$. Similarly (36) may be rewritten as:

$$\begin{pmatrix} c_2 & s_2 & 0 \\ s_2 & -c_2 & 0 \\ 0 & 0 & 1 \end{pmatrix} \tilde{\mathbf{b}} = \begin{pmatrix} 1 & 0 & 0 \\ 0 & -\lambda_2 & \mu_2 \\ 0 & \mu_2 & \lambda_2 \end{pmatrix} \begin{pmatrix} c_3 & s_3 & 0 \\ s_3 & -c_3 & 0 \\ 0 & 0 & 1 \end{pmatrix} \tilde{\mathbf{r}}, \tag{38}$$

where $\tilde{\mathbf{b}} = \begin{pmatrix} b_1 \\ b_2 \\ b_3 \end{pmatrix}$ and $\tilde{\mathbf{r}} = \begin{pmatrix} r_1 \\ r_2 \\ r_3 \end{pmatrix}$.

Henceforth we will refer to (37) and (38) by the vectors $\tilde{\mathbf{p}}$ and $\tilde{\mathbf{n}}$ respectively. By this we mean

$$\tilde{\mathbf{p}} = \begin{pmatrix} c_2 & s_2 & 0 \\ s_2 & -c_2 & 0 \\ 0 & 0 & 1 \end{pmatrix} \tilde{\mathbf{h}} =$$

$$\begin{pmatrix} 1 & 0 & 0 \\ 0 & -\lambda_2 & \mu_2 \\ 0 & \mu_2 & \lambda_2 \end{pmatrix} \begin{pmatrix} c_3 & s_3 & 0 \\ s_3 & -c_3 & 0 \\ 0 & 0 & 1 \end{pmatrix} \tilde{\mathbf{f}} + \begin{pmatrix} a_2 \\ 0 \\ d_2 \end{pmatrix},$$

and

$$\tilde{\mathbf{n}} = \begin{pmatrix} c_2 & s_2 & 0 \\ s_2 & -c_2 & 0 \\ 0 & 0 & 1 \end{pmatrix} \tilde{\mathbf{b}} = \begin{pmatrix} 1 & 0 & 0 \\ 0 & -\lambda_2 & \mu_2 \\ 0 & \mu_2 & \lambda_2 \end{pmatrix} \begin{pmatrix} c_3 & s_3 & 0 \\ s_3 & -c_3 & 0 \\ 0 & 0 & 1 \end{pmatrix} \tilde{\mathbf{r}}.$$

We now proceed to eliminate four of the five variables in $\tilde{\mathbf{p}}$ and $\tilde{\mathbf{n}}$ by exploiting the structure of the ideal generated by the component equations of $\tilde{\mathbf{p}}$ and $\tilde{\mathbf{n}}$. We know from Properties 1-4 of the previous section that $\tilde{\mathbf{p}} \cdot \tilde{\mathbf{p}}$, $\tilde{\mathbf{p}} \cdot \tilde{\mathbf{n}}$, $\tilde{\mathbf{p}} \times \tilde{\mathbf{n}}$, and $(\tilde{\mathbf{p}} \cdot \tilde{\mathbf{p}})\tilde{\mathbf{n}} - (2\tilde{\mathbf{p}} \cdot \tilde{\mathbf{n}})\tilde{\mathbf{p}}$, all have the same structure as $\tilde{\mathbf{p}}$ and $\tilde{\mathbf{n}}$ viz. they are all of degree 1 or less in $\sin\theta_i, \cos\theta_i$, $i = 1, \ldots, 5$. Consequently we have the following set of fourteen equations all of which have the same set of power products:[4]

Vector	No. of Scalar Equations
$\tilde{\mathbf{p}}$	3
$\tilde{\mathbf{n}}$	3
$\tilde{\mathbf{p}} \cdot \tilde{\mathbf{p}}$	1
$\tilde{\mathbf{p}} \cdot \tilde{\mathbf{n}}$	1
$\tilde{\mathbf{p}} \times \tilde{\mathbf{n}}$	3
$(\tilde{\mathbf{p}} \cdot \tilde{\mathbf{p}})\tilde{\mathbf{n}} - (2\tilde{\mathbf{p}} \cdot \tilde{\mathbf{n}})\tilde{\mathbf{p}}$	3
	(Total) 14

These fourteen equations are linearly independent. They may be written in matrix form as:

$$(P)\begin{pmatrix} s_4s_5 \\ s_4c_5 \\ c_4s_5 \\ c_4c_5 \\ s_4 \\ c_4 \\ s_5 \\ c_5 \\ 1 \end{pmatrix} = (Q)\begin{pmatrix} s_1s_2 \\ s_1c_2 \\ c_1s_2 \\ c_1c_2 \\ s_1 \\ c_1 \\ s_2 \\ c_2 \end{pmatrix}, \tag{39}$$

where P is a 14×9 matrix, the entries of which are linear combinations of $s_3, c_3, 1$, and Q is a 14×8 matrix, the entries of which are constants. We use any eight of the fourteen equations in (39) to solve for the eight right-hand side terms containing θ_1 and θ_2 in terms of the left-hand side which is a function of θ_3, θ_4 and θ_5. We use these to eliminate terms containing θ_1 and θ_2 from the remaining six equations, which then take the form:

$$(\Sigma)\begin{pmatrix} s_4s_5 \\ s_4c_5 \\ c_4s_5 \\ c_4c_5 \\ s_4 \\ c_4 \\ s_5 \\ c_5 \\ 1 \end{pmatrix} = 0, \tag{40}$$

[4]By power products we mean "terms," e.g. the power products of the polynomial $5x^2y + 3xz + 9y^2 + 4z = 0$ are x^2y, xz, y^2 and z.

where Σ is a 6×9 matrix, the entries of which are linear combinations of $s_3, c_3, 1$. We make the following substitutions in (40):

$$s_4 \leftarrow \frac{2x_4}{1+x_4^2}, c_4 \leftarrow \frac{1-x_4^2}{1+x_4^2}, s_5 \leftarrow \frac{2x_5}{1+x_5^2}, c_5 \leftarrow \frac{1-x_5^2}{1+x_5^2},$$

where $x_4 = tan(\frac{\theta_4}{2}), x_5 = tan(\frac{\theta_5}{2})$.

We then multiply each equation by $(1+x_4^2)$ and $(1+x_5^2)$ to clear denominators. (40) then takes the form

$$(\Sigma') \begin{pmatrix} x_4^2x_5^2 \\ x_4^2x_5 \\ x_4^2 \\ x_4x_5^2 \\ x_4x_5 \\ x_4 \\ x_5^2 \\ x_5 \\ 1 \end{pmatrix} = 0, \tag{41}$$

where Σ' is a 6×9 matrix, the entries of which are linear combinations of $s_3, c_3, 1$. We make the following substitutions in (41):

$$s_3 \leftarrow \frac{2x_3}{1+x_3^2}, c_3 \leftarrow \frac{1-x_3^2}{1+x_3^2}.$$

We multiply the first four scalar equations in (41) by $(1+x_3^2)$ to clear denominators. The resulting equation is of the form

$$(\Sigma'') \begin{pmatrix} x_4^2x_5^2 \\ x_4^2x_5 \\ x_4^2 \\ x_4x_5^2 \\ x_4x_5 \\ x_4 \\ x_4 \\ x_5^2 \\ x_5 \\ 1 \end{pmatrix} = 0, \tag{42}$$

where Σ'' is a 6×9 matrix. The entries in the first four rows of Σ'' are quadratic polynomials in x_3. The entries in the last two rows are rational functions of x_3, the numerators being quadratic polynomials in x_3, the denominators being $(1+x_3^2)$. It is noteworthy that the determinant of the 6×6 array comprised of any set of six columns of Σ'' is always an 8th degree polynomial *and not a rational function.* This fact is proved rigorously in Appendix 2.

We now eliminate x_4 and x_5 dialytically as follows : multiplying (42) by x_4, we get the following equation:

$$(\Sigma'') \begin{pmatrix} x_4^3 x_5^2 \\ x_4^3 x_5 \\ x_4^3 \\ x_4^2 x_5^2 \\ x_4^2 x_5 \\ x_4^2 \\ x_4 x_5^2 \\ x_4 x_5 \\ x_4 \end{pmatrix} = 0. \tag{43}$$

(42) and (43) taken together may be written in matrix form as:

$$\begin{pmatrix} \Sigma'' & 0 \\ 0 & \Sigma'' \end{pmatrix} \begin{pmatrix} x_4^3 x_5^2 \\ x_4^3 x_5 \\ x_4^3 \\ x_4^2 x_5^2 \\ x_4^2 x_5 \\ x_4^2 \\ x_4 x_5^2 \\ x_4 x_5 \\ x_4 \\ x_5^2 \\ x_5 \\ 1 \end{pmatrix} = 0. \tag{44}$$

The determinant of the coefficient matrix is a 16th degree polynomial in x_3. The roots of this polynomial give the values of x_3 corresponding to the sixteen solutions of the inverse kinematics problem. For each value of x_3 thus obtained, the corresponding value of θ_3 may be computed using the formula $\theta_3 = 2tan^{-1}x_3$. It so happens that a 6R manipulator can place its hand in a specified position and orientation in sixteen different ways. Therefore the above procedure gives no extraneous roots.

For each value of θ_3, we may compute the remaining joint variables as follows : we substitute the numerical value of θ_3 in the coefficient matrix of (44). We then use eleven independent members of (44) to solve for the eleven terms $x_4^3x_5^2$, $x_4^3x_5$, x_4^3, $x_4^2x_5^2$, $x_4^2x_5$, x_4^2, $x_4x_5^2$, x_4x_5, x_4, x_5^2, x_5. The numerical values of x_4 and x_5 may be used to compute θ_4 and θ_5. We then substitute numerical values for θ_3, θ_4, and θ_5 in (39). We use eight linearly independent members of the resulting equation to solve for s_1s_2, s_1c_2, c_1s_2, c_1c_2, s_1, c_1, s_2 and c_2. We then use the numerical values of s_1 and c_1 to obtain a unique value for θ_1. Similarly, θ_2 may be computed using the numerical values of s_2 and c_2. Finally, substituting values for θ_1, θ_2, θ_3, θ_4, and θ_5 in the $(1,1)$ and $(2,1)$ elements of the following equation

$${}^{7}_{6}\mathbf{T} = {}^{6}_{5}\mathbf{T}^{-1}\,{}^{5}_{4}\mathbf{T}^{-1}\,{}^{4}_{3}\mathbf{T}^{-1}\,{}^{3}_{2}\mathbf{T}^{-1}\,{}^{2}_{1}\mathbf{T}^{-1}\,{}^{hand}_{1}\mathbf{T}, \tag{45}$$

yields two linear equations in s_6 and c_6. After solving for s_6 and c_6 we may use their values to determine a unique value for θ_6.

6. Brief history of manipulator kinematics. The use of coordinate systems and transformation matrices to model linkages was introduced by DENAVIT and HARTENBERG[1955]. The first systematic study of the inverse kinematics of spatial series manipulators with revolute and prismatic joints was performed by PIEPER[1968]. He presented a classification of manipulators based on the degree of difficulty of their inverse kinematics. He worked out closed-form solutions for various manipulators. Pieper recognized that an algebraic solution for the 6R would be quite difficult. Therefore he suggested a numerical solution. In the years following the appearance of Pieper's thesis, algebraic solutions to the inverse kinematics problem for most series manipulators with six or fewer degrees of freedom were presented. However the 6R remained unsolved. ROTH, RASTEGAR and SCHEINMAN[1973] showed that in general this problem has at most thirty-two solutions, i.e. there are at most thirty-two distinct configurations of the manipulator, in which its hand is at the desired goal position. Their proof was based on arguments from synthetic geometry and was nonconstructive. ALBALA and ANGELES[1979], using indicial notation, presented the first constructive solution to this problem. Their solution was in the form of a 12×12 determinant the entries of which were quartic polynomials in the tangent of the half-angle of one of the joint variables. DUFFY and CRANE[1980] later provided a different solution by combining spherical trigonometry and dialytical elimination to obtain a 32nd degree polynomial in the tangent of the half-angle of one of the joint variables. Their 32nd degree polynomial always yielded extraneous or false roots in addition to the desired values. TSAI and MORGAN[1985] used homotopy continuation to solve the inverse kinematics problem and found only sixteen solutions for various 6R manipulators of different geometries. They therefore conjectured that this problem has at most sixteen solutions. Recently, LEE and LIANG[1988a and 1988b], used extensions of Duffy's method to obtain by dialytical elimination a 16th degree polynomial in the tangent of the half-angle of one of the joint variables, thus confirming the conjecture of TSAI and MORGAN[1985]. We believe that the solution presented in this manuscript is much simpler than the procedure proposed by Lee and Liang.

A1. Appendix 1.

A1.1 LEMMA. *The equations obtained by equating entries from the left and right-hand sides of (20) are of degree 1 or less in* $\sin\theta_i, \cos\theta_i, i = 1, 2, \ldots, n$.

By this we mean that for any given integer i, the highest power of $\sin\theta_i, \cos\theta_i$ appearing in these equations is less than or equal to 1.

A1.2 PROOF. The entries of ${}_{i}^{i+1}\mathbf{T}$ are of degree 1 or less in $\sin\theta_i, \cos\theta_i$. The lemma follows naturally from this fact.

REMARK. By A1.1 (23), (24), (25), and (26) are of degree 1 or less in $\sin\theta_i, \cos\theta_i, i = 1, 2, \ldots, n$.

A1.3 LEMMA. *The entries of* ${}^{n+1}_{1}\mathbf{T}^{-1}$ *are of degree 1 or less in* $\sin\theta_i, \cos\theta_i, i =$ $1, 2, \ldots, n$.

A1.4 PROOF. By definition

$$ {}^{n+1}_{1}\mathbf{T} = {}^{2}_{1}\mathbf{T}{}^{3}_{2}\mathbf{T}\ldots{}^{n}_{n-1}\mathbf{T}{}^{n+1}_{n}\mathbf{T}. \tag{46} $$

Therefore

$$ {}^{n+1}_{1}\mathbf{T}^{-1} = {}^{n+1}_{n}\mathbf{T}^{-1}{}^{n}_{n-1}\mathbf{T}^{-1}{}^{n-1}_{n-2}\mathbf{T}^{-1}\ldots{}^{2}_{1}\mathbf{T}^{-1}. \tag{47} $$

The reader may verify that for any i

$$ {}^{i+1}_{i}\mathbf{T}^{-1} = \begin{pmatrix} \cos\theta_i & \sin\theta_i & 0 & -a_i \\ -\sin\theta_i\cos\alpha_i & \cos\theta_i\cos\alpha_i & \sin\alpha_i & -d_i\sin\alpha_i \\ \sin\theta_i\sin\alpha_i & -\cos\theta_i\sin\alpha_i & \cos\alpha_i & -d_i\cos\alpha_i \\ 0 & 0 & 0 & 1 \end{pmatrix}. \tag{48} $$

Since each entry of ${}^{i+1}_{i}\mathbf{T}^{-1}$ is of degree 1 or less in $\sin\theta_i, \cos\theta_i$, each matrix on the right-hand side of (47) satisfies this property and the lemma follows.

A1.5 LEMMA. *Each entry of the vector* ${}^{n}_{1}\mathbf{R}^{T}{}^{n}_{1}\mathbf{p}$ *is of degree 1 or less in* $\sin\theta_i$, $\cos\theta_i$, $i = 1, 2, \ldots, n-1$.

A1.6 PROOF. Using the notation presented in Section 4 we may write ${}^{n}_{1}\mathbf{T}$ as $\begin{pmatrix} {}^{n}_{1}\mathbf{R} & {}^{n}_{1}\mathbf{p} \\ 0 & 1 \end{pmatrix}$. In this notation,

$$ {}^{n}_{1}\mathbf{T}^{-1} = \begin{pmatrix} {}^{n}_{1}\mathbf{R}^{T} & -{}^{n}_{1}\mathbf{R}^{T}{}^{n}_{1}\mathbf{p} \\ 0 & 1 \end{pmatrix}. \tag{49} $$

(49) may be verified very easily by seeing that ${}^{n}_{1}\mathbf{T}^{-1}{}^{n}_{1}\mathbf{T}$ is equal to I. The lemma follows from (49) and A1.3.

A1.7 LEMMA. *Let* $\mathbf{v}_1, \mathbf{v}_2$ *be arbitrary vectors. Then* $({}^{i}_{j}\mathbf{R}\mathbf{v}_1) \times ({}^{i}_{j}\mathbf{R}\mathbf{v}_2) = {}^{i}_{j}\mathbf{R}(\mathbf{v}_1 \times \mathbf{v}_2)$, $\forall\ i, j$.

A1.8 PROOF. Let $\mathbf{v}_1 = \begin{pmatrix} v_{11} & v_{12} & v_{13} \end{pmatrix}^T$, $\mathbf{v}_2 = \begin{pmatrix} v_{21} & v_{22} & v_{23} \end{pmatrix}^T$.

$$ {}^{i}_{j}\mathbf{R}\mathbf{v}_1 = \begin{pmatrix} {}^{i}_{j}\mathbf{l} & {}^{i}_{j}\mathbf{m} & {}^{i}_{j}\mathbf{n} \end{pmatrix}\mathbf{v}_1 = {}^{i}_{j}\mathbf{l}v_{11} + {}^{i}_{j}\mathbf{m}v_{12} + {}^{i}_{j}\mathbf{n}v_{13}. \tag{50} $$

Similarly

$$ {}^{i}_{j}\mathbf{R}\mathbf{v}_2 = {}^{i}_{j}\mathbf{l}v_{21} + {}^{i}_{j}\mathbf{m}v_{22} + {}^{i}_{j}\mathbf{n}v_{23}. \tag{51} $$

From (50) and (51), it follows that

$$ \begin{aligned} ({}^{i}_{j}\mathbf{R}\mathbf{v}_1) \times ({}^{i}_{j}\mathbf{R}\mathbf{v}_2) &= ({}^{i}_{j}\mathbf{l} \times {}^{i}_{j}\mathbf{l})v_{11}v_{21} + ({}^{i}_{j}\mathbf{l} \times {}^{i}_{j}\mathbf{m})v_{11}v_{22} + ({}^{i}_{j}\mathbf{l} \times {}^{i}_{j}\mathbf{n})v_{11}v_{23} \\ +({}^{i}_{j}\mathbf{m} \times {}^{i}_{j}\mathbf{l})v_{12}v_{21} &+ ({}^{i}_{j}\mathbf{m} \times {}^{i}_{j}\mathbf{m})v_{12}v_{22} + ({}^{i}_{j}\mathbf{m} \times {}^{i}_{j}\mathbf{n})v_{12}v_{23} \\ +({}^{i}_{j}\mathbf{n} \times {}^{i}_{j}\mathbf{l})v_{13}v_{21} &+ ({}^{i}_{j}\mathbf{n} \times {}^{i}_{j}\mathbf{m})v_{13}v_{22} + ({}^{i}_{j}\mathbf{n} \times {}^{i}_{j}\mathbf{n})v_{13}v_{23}. \end{aligned} \tag{52} $$

Since ${}^i_j\mathbf{R}$ is orthogonal, ${}^i_j\mathbf{l} \times {}^i_j\mathbf{m} = {}^i_j\mathbf{n}$, ${}^i_j\mathbf{m} \times {}^i_j\mathbf{n} = {}^i_j\mathbf{l}$, ${}^i_j\mathbf{n} \times {}^i_j\mathbf{l} = {}^i_j\mathbf{m}$. Using these relations (52) simplifies to

$$\begin{aligned}({}^i_j\mathbf{R}\mathbf{v}_1) \times ({}^i_j\mathbf{R}\mathbf{v}_2) &= {}^i_j\mathbf{l}(v_{12}v_{23} - v_{13}v_{22}) \\ &\quad + {}^i_j\mathbf{m}(v_{13}v_{21} - v_{11}v_{23}) \\ &\quad + {}^i_j\mathbf{n}(v_{11}v_{22} - v_{12}v_{21}), \\ &= {}^i_j\mathbf{R}(\mathbf{v}_1 \times \mathbf{v}_2). \end{aligned} \tag{53}$$

A1.9 PROPERTY 1. *${}^{n+1}_1\mathbf{p} \cdot {}^{n+1}_1\mathbf{p}$ is of degree 1 or less in $\sin\theta_i, \cos\theta_i, i = 1, 2, \ldots, n$.*

A1.10 PROOF. For $n = 1$, the term ${}^{n+1}_1\mathbf{p} \cdot {}^{n+1}_1\mathbf{p} = {}^2_1\mathbf{p} \cdot {}^2_1\mathbf{p} = a_1^2 + d_1^2$, a constant. Therefore Property 1 is true for $n = 1$. Assume it is true for n equal to some integer $p - 1$, greater than 1. Then for n equal to p, the function ${}^{n+1}_1\mathbf{p} \cdot {}^{n+1}_1\mathbf{p}$ may be written as follows (using (26)):

$$\begin{aligned}{}^{p+1}_1\mathbf{p} \cdot {}^{p+1}_1\mathbf{p} &= \{{}^p_1\mathbf{R}{}^{p+1}_p\mathbf{p} + {}^p_1\mathbf{p}\} \cdot \{{}^p_1\mathbf{R}{}^{p+1}_p\mathbf{p} + {}^p_1\mathbf{p}\}, \\ &= {}^p_1\mathbf{p} \cdot {}^p_1\mathbf{p} + ({}^{p+1}_p\mathbf{p}) \cdot ({}^{p+1}_p\mathbf{p}) + 2{}^{p+1}_p\mathbf{p}^T({}^p_1\mathbf{R}^T{}^p_1\mathbf{p}). \end{aligned} \tag{54}$$

The term ${}^{p+1}_p\mathbf{p} \cdot {}^{p+1}_p\mathbf{p}$ is equal to $a_p^2 + d_p^2$. By A1.5 the entries of ${}^p_1\mathbf{R}^T{}^p_1\mathbf{p}$ are of degree 1 or less in $\sin\theta_i, \cos\theta_i, i = 1, 2, \ldots, p-1$. The vector ${}^{p+1}_p\mathbf{p}$ is by definition of degree 1 or less in $\sin\theta_p, \cos\theta_p$. Therefore the term $2{}^{p+1}_1\mathbf{p}^T({}^p_1\mathbf{R}^T{}^p_1\mathbf{p})$ is of degree 1 or less in $\sin\theta_i, \cos\theta_i, i = 1, 2, \ldots, p$. Consequently (54) may be written as

$${}^{p+1}_1\mathbf{p} \cdot {}^{p+1}_1\mathbf{p} = {}^p_1\mathbf{p} \cdot {}^p_1\mathbf{p} + \textit{terms of degree } 1 \textit{ or less in } \sin\theta_i, \cos\theta_i, i = 1, 2, \ldots, p. \tag{55}$$

By induction on n, Property 1 is true for $n \geq 1$.

A1.11 PROPERTY 2. *Each of the functions $({}^{n+1}_1\mathbf{p} \cdot {}^{n+1}_1\mathbf{l})$, $({}^{n+1}_1\mathbf{p} \cdot {}^{n+1}_1\mathbf{m})$, $({}^{n+1}_1\mathbf{p} \cdot {}^{n+1}_1\mathbf{n})$, is of degree 1 or less in $\sin\theta_i, \cos\theta_i, i = 1, 2, \ldots, n$.*

A1.12 PROOF. Substituting from (23) and (26)

$$\begin{aligned}{}^{n+1}_1\mathbf{p} \cdot {}^{n+1}_1\mathbf{l} &= ({}^n_1\mathbf{R}{}^{n+1}_n\mathbf{p} + {}^n_1\mathbf{p}) \cdot ({}^n_1\mathbf{R}{}^{n+1}_n\mathbf{l}), \\ &= ({}^{n+1}_n\mathbf{p}) \cdot ({}^{n+1}_n\mathbf{l}) + {}^{n+1}_n\mathbf{l}^T({}^n_1\mathbf{R}^T{}^n_1\mathbf{p}). \end{aligned} \tag{56}$$

The term ${}^{n+1}_n\mathbf{p} \cdot {}^{n+1}_n\mathbf{l}$ is equal to a_n, a constant. ${}^{n+1}_n\mathbf{l}$ is by definition of degree 1 or less in $\sin\theta_n$, $\cos\theta_n$. By A1.5, ${}^n_1\mathbf{R}^T{}^n_1\mathbf{p}$ is of degree 1 or less in $\sin\theta_i, \cos\theta_i, i = 1, 2, \ldots, n-1$. Therefore the term ${}^{n+1}_n\mathbf{l}^T({}^n_1\mathbf{R}^T{}^n_1\mathbf{p})$ is of degree 1 or less in $\sin\theta_i, \cos\theta_i, i = 1, 2, \ldots, n$. Consequently Property 2 is true for the function ${}^{n+1}_1\mathbf{p} \cdot {}^{n+1}_1\mathbf{l}$. Proofs for the functions ${}^{n+1}_1\mathbf{p} \cdot {}^{n+1}_1\mathbf{m}$ and ${}^{n+1}_1\mathbf{p} \cdot {}^{n+1}_1\mathbf{n}$ are along the same lines.

A1.13 PROPERTY 3. *The entries of the vectors ${}^{n+1}_1\mathbf{p} \times {}^{n+1}_1\mathbf{l}$, ${}^{n+1}_1\mathbf{p} \times {}^{n+1}_1\mathbf{m}$, ${}^{n+1}_1\mathbf{p} \times {}^{n+1}_1\mathbf{n}$ are of degree 1 or less in $\sin\theta_i, \cos\theta_i, i = 1, 2, \ldots, n$.*

A1.14 PROOF. For $n = 1$, the above vectors are:

$$ {}^{2}_{1}\mathbf{p} \times {}^{2}_{1}\mathbf{l} = \begin{pmatrix} -d_1 \sin\theta_1 \\ d_1 \cos\theta_1 \\ 0 \end{pmatrix}, \tag{57} $$

$$ {}^{2}_{1}\mathbf{p} \times {}^{2}_{1}\mathbf{m} = \begin{pmatrix} a_1 \sin\theta_1 \sin\alpha_1 - d_1 \cos\theta_1 \cos\alpha_1 \\ -d_1 \sin\theta_1 \cos\alpha_1 - a_1 \cos\theta_1 \sin\alpha_1 \\ a_1 \cos\alpha_1 \end{pmatrix}, \tag{58} $$

$$ {}^{2}_{1}\mathbf{p} \times {}^{2}_{1}\mathbf{n} = \begin{pmatrix} a_1 \sin\theta_1 \cos\alpha_1 + d_1 \cos\theta_1 \sin\alpha_1 \\ d_1 \sin\theta_1 \sin\alpha_1 - a_1 \cos\theta_1 \cos\alpha_1 \\ -a_1 \sin\alpha_1 \end{pmatrix}. \tag{59} $$

Clearly Property 3 is true when n is equal to 1. Assume that it is true when n is equal to some integer $p-1$, greater than 1. For n equal to p, the vector ${}^{p+1}_{1}\mathbf{p} \times {}^{p+1}_{1}\mathbf{l}$ may be written as follows (using (23) and (26)):

$$ \begin{aligned} {}^{p+1}_{1}\mathbf{p} \times {}^{p+1}_{1}\mathbf{l} &= ({}^{p}_{1}\mathbf{R}^{p+1}_{p}\mathbf{p} + {}^{p}_{1}\mathbf{p}) \times ({}^{p}_{1}\mathbf{R}^{p+1}_{p}\mathbf{l}), \\ &= {}^{p}_{1}\mathbf{R}({}^{p+1}_{p}\mathbf{p} \times {}^{p+1}_{p}\mathbf{l}) \\ &+ \left({}^{p}_{1}\mathbf{p} \times {}^{p}_{1}\mathbf{l} \quad {}^{p}_{1}\mathbf{p} \times {}^{p}_{1}\mathbf{m} \quad {}^{p}_{1}\mathbf{p} \times {}^{p}_{1}\mathbf{n} \right) {}^{p+1}_{p}\mathbf{l}. \end{aligned} \tag{60} $$

Similarly

$$ {}^{p+1}_{1}\mathbf{p} \times {}^{p+1}_{1}\mathbf{m} = {}^{p}_{1}\mathbf{R}({}^{p+1}_{p}\mathbf{p} \times {}^{p+1}_{p}\mathbf{m}) + \left({}^{p}_{1}\mathbf{p} \times {}^{p}_{1}\mathbf{l} \quad {}^{p}_{1}\mathbf{p} \times {}^{p}_{1}\mathbf{m} \quad {}^{p}_{1}\mathbf{p} \times {}^{p}_{1}\mathbf{n} \right) {}^{p+1}_{p}\mathbf{m}. \tag{61} $$

$$ {}^{p+1}_{1}\mathbf{p} \times {}^{p+1}_{1}\mathbf{n} = {}^{p}_{1}\mathbf{R}({}^{p+1}_{p}\mathbf{p} \times {}^{p+1}_{p}\mathbf{n}) + \left({}^{p}_{1}\mathbf{p} \times {}^{p}_{1}\mathbf{l} \quad {}^{p}_{1}\mathbf{p} \times {}^{p}_{1}\mathbf{m} \quad {}^{p}_{1}\mathbf{p} \times {}^{p}_{1}\mathbf{n} \right) {}^{p+1}_{p}\mathbf{n}. \tag{62} $$

Each entry of the matrix $\left({}^{p}_{1}\mathbf{p} \times {}^{p}_{1}\mathbf{l} \quad {}^{p}_{1}\mathbf{p} \times {}^{p}_{1}\mathbf{m} \quad {}^{p}_{1}\mathbf{p} \times {}^{p}_{1}\mathbf{n} \right)$ is of degree 1 or less in $\sin\theta_i, \cos\theta_i, i = 1, 2, \ldots, p-1$, because by assumption, Property 3 is true for n equal to $p-1$. Therefore the vectors $\left({}^{p}_{1}\mathbf{p} \times {}^{p}_{1}\mathbf{l} \quad {}^{p}_{1}\mathbf{p} \times {}^{p}_{1}\mathbf{m} \quad {}^{p}_{1}\mathbf{p} \times {}^{p}_{1}\mathbf{n} \right) {}^{p+1}_{p}\mathbf{l}$, $\left({}^{p}_{1}\mathbf{p} \times {}^{p}_{1}\mathbf{l} \quad {}^{p}_{1}\mathbf{p} \times {}^{p}_{1}\mathbf{m} \quad {}^{p}_{1}\mathbf{p} \times {}^{p}_{1}\mathbf{n} \right) {}^{p+1}_{p}\mathbf{m}$, and $\left({}^{p}_{1}\mathbf{p} \times {}^{p}_{1}\mathbf{l} \quad {}^{p}_{1}\mathbf{p} \times {}^{p}_{1}\mathbf{m} \quad {}^{p}_{1}\mathbf{p} \times {}^{p}_{1}\mathbf{n} \right) {}^{p+1}_{p}\mathbf{n}$, appearing in (60), (61), and (62) are of degree 1 or less in $\sin\theta_i, \cos\theta_i, i = 1, 2, \ldots, p$. The reader may verify that the entries of the vectors ${}^{p+1}_{p}\mathbf{p} \times {}^{p+1}_{p}\mathbf{l}$, ${}^{p+1}_{p}\mathbf{p} \times {}^{p+1}_{p}\mathbf{m}$, and ${}^{p+1}_{p}\mathbf{p} \times {}^{p+1}_{p}\mathbf{n}$, are of degree 1 or less in $\sin\theta_p, \cos\theta_p$. Therefore it follows that the vectors on the right-hand side of (60), (61), and (62) are of degree 1 or less in $\sin\theta_i, \cos\theta_i, i = 1, 2, \ldots, p$. By induction on n, Property 3 is true for $n \geq 1$.

A1.15 PROPERTY 4. *The functions* $({}^{n+1}_{1}\mathbf{p} \cdot {}^{n+1}_{1}\mathbf{p}){}^{n+1}_{1}\mathbf{l} - (2\,{}^{n+1}_{1}\mathbf{p} \cdot {}^{n+1}_{1}\mathbf{l}){}^{n+1}_{1}\mathbf{p}$, $({}^{n+1}_{1}\mathbf{p} \cdot {}^{n+1}_{1}\mathbf{p}){}^{n+1}_{1}\mathbf{m} - (2\,{}^{n+1}_{1}\mathbf{p} \cdot {}^{n+1}_{1}\mathbf{m}){}^{n+1}_{1}\mathbf{p}$, *and* $({}^{n+1}_{1}\mathbf{p} \cdot {}^{n+1}_{1}\mathbf{p}){}^{n+1}_{1}\mathbf{n} - (2\,{}^{n+1}_{1}\mathbf{p} \cdot {}^{n+1}_{1}\mathbf{n}){}^{n+1}_{1}\mathbf{p}$ *are of degree 1 or less in* $\sin\theta_i, \cos\theta_i, i = 1, 2, \ldots, n$.

A1.16 PROOF. Setting n equal to 1 in the above functions, we get

$$ ({}^{2}_{1}\mathbf{p} \cdot {}^{2}_{1}\mathbf{p}){}^{2}_{1}\mathbf{l} - (2\,{}^{2}_{1}\mathbf{p} \cdot {}^{2}_{1}\mathbf{l}){}^{2}_{1}\mathbf{p} = (a_1^2 + d_1^2){}^{2}_{1}\mathbf{l} - 2a_1({}^{2}_{1}\mathbf{p}), \tag{63} $$

$$ ({}^{2}_{1}\mathbf{p} \cdot {}^{2}_{1}\mathbf{p}){}^{2}_{1}\mathbf{m} - (2\,{}^{2}_{1}\mathbf{p} \cdot {}^{2}_{1}\mathbf{m}){}^{2}_{1}\mathbf{p} = (a_1^2 + d_1^2){}^{2}_{1}\mathbf{m} - 2d_1 \sin\alpha_1({}^{2}_{1}\mathbf{p}), \tag{64} $$

$$({}^2_1\mathbf{p}\cdot{}^2_1\mathbf{p}){}^2_1\mathbf{n} - (2{}^2_1\mathbf{p}\cdot{}^2_1\mathbf{n}){}^2_1\mathbf{p} = (a_1^2+d_1^2){}^2_1\mathbf{n} - 2d_1\cos\alpha_1({}^2_1\mathbf{p}). \tag{65}$$

Clearly Property 4 is true when n is equal to 1. Let us assume that it is true when n is equal to some integer $p-1$, greater than 1.

For n equal to p, we get (using (23), (26), (54), and (56)):

$$\begin{aligned}
&({}^{p+1}_{1}\mathbf{p}\cdot{}^{p+1}_{1}\mathbf{p}){}^{p+1}_{1}\mathbf{l} - (2{}^{p+1}_{1}\mathbf{p}\cdot{}^{p+1}_{1}\mathbf{l}){}^{p+1}_{1}\mathbf{p} = \\
&\{{}^{p+1}_{p}\mathbf{p}\cdot{}^{p+1}_{p}\mathbf{p} + {}^{p}_{1}\mathbf{p}\cdot{}^{p}_{1}\mathbf{p} + 2({}^{p}_{1}\mathbf{R}{}^{p+1}_{p}\mathbf{p})\cdot{}^{p}_{1}\mathbf{p}\}{}^{p}_{1}\mathbf{R}{}^{p+1}_{p}\mathbf{l} \\
&-2\{{}^{p+1}_{p}\mathbf{p}\cdot{}^{p+1}_{p}\mathbf{l} + {}^{p}_{1}\mathbf{p}\cdot({}^{p}_{1}\mathbf{R}{}^{p+1}_{p}\mathbf{l})\}\{{}^{p}_{1}\mathbf{R}{}^{p+1}_{p}\mathbf{p} + {}^{p}_{1}\mathbf{p}\}.
\end{aligned} \tag{66}$$

After some expansion and simplification, (66) may be written as

$$\begin{aligned}
&({}^{p+1}_{1}\mathbf{p}\cdot{}^{p+1}_{1}\mathbf{p}){}^{p+1}_{1}\mathbf{l} - (2{}^{p+1}_{1}\mathbf{p}\cdot{}^{p+1}_{1}\mathbf{l}){}^{p+1}_{1}\mathbf{p} = \\
&{}^{p}_{1}\mathbf{R}\{({}^{p+1}_{p}\mathbf{p}\cdot{}^{p+1}_{p}\mathbf{p}){}^{p+1}_{p}\mathbf{l} - (2{}^{p+1}_{p}\mathbf{p}\cdot{}^{p+1}_{p}\mathbf{l}){}^{p+1}_{p}\mathbf{p}\} \\
&+\begin{pmatrix} \{({}^{p}_{1}\mathbf{p}\cdot{}^{p}_{1}\mathbf{p}){}^{p}_{1}\mathbf{l} - (2{}^{p}_{1}\mathbf{p}\cdot{}^{p}_{1}\mathbf{l}){}^{p}_{1}\mathbf{p}\}^T \\ \{({}^{p}_{1}\mathbf{p}\cdot{}^{p}_{1}\mathbf{p}){}^{p}_{1}\mathbf{m} - (2{}^{p}_{1}\mathbf{p}\cdot{}^{p}_{1}\mathbf{m}){}^{p}_{1}\mathbf{p}\}^T \\ \{({}^{p}_{1}\mathbf{p}\cdot{}^{p}_{1}\mathbf{p}){}^{p}_{1}\mathbf{n} - (2{}^{p}_{1}\mathbf{p}\cdot{}^{p}_{1}\mathbf{n}){}^{p}_{1}\mathbf{p}\}^T \end{pmatrix}^T {}^{p+1}_{p}\mathbf{l} \\
&+2{}^{p}_{1}\mathbf{p}\times{}^{p}_{1}\mathbf{R}({}^{p+1}_{p}\mathbf{l}\times{}^{p+1}_{p}\mathbf{p}) - 2({}^{p+1}_{p}\mathbf{p}\cdot{}^{p+1}_{p}\mathbf{l}){}^{p}_{1}\mathbf{p}.
\end{aligned} \tag{67}$$

The term $2{}^{p}_{1}\mathbf{p}\times{}^{p}_{1}\mathbf{R}({}^{p+1}_{p}\mathbf{l}\times{}^{p+1}_{p}\mathbf{p})$ is equal to $2\begin{pmatrix} {}^{p}_{1}\mathbf{p}\times{}^{p}_{1}\mathbf{l} & {}^{p}_{1}\mathbf{p}\times{}^{p}_{1}\mathbf{m} & {}^{p}_{1}\mathbf{p}\times{}^{p}_{1}\mathbf{n}\end{pmatrix}$ $({}^{p+1}_{p}\mathbf{l}\times{}^{p+1}_{p}\mathbf{p})$ and is clearly of degree 1 or less in $\sin\theta_i, \cos\theta_i, i = 1,2,\ldots,p$, by A1.13 (Property 3.) The term $-2{}^{p}_{1}\mathbf{p}({}^{p+1}_{p}\mathbf{p}\cdot{}^{p+1}_{p}\mathbf{l})$ is also of degree 1 or less in $\sin\theta_i, \cos\theta_i, i = 1,2,\ldots,p$. The term $({}^{p+1}_{p}\mathbf{p}\cdot{}^{p+1}_{p}\mathbf{p}){}^{p+1}_{p}\mathbf{l} - (2{}^{p+1}_{p}\mathbf{p}\cdot{}^{p+1}_{p}\mathbf{l}){}^{p+1}_{p}\mathbf{p}$ is equal to $(a_p^2+d_p^2){}^{p+1}_{p}\mathbf{l} - (2a_p){}^{p+1}_{p}\mathbf{p}$, which is clearly of degree 1 or less in $\sin\theta_p, \cos\theta_p$. Therefore the term ${}^{p}_{1}\mathbf{R}(({}^{p+1}_{p}\mathbf{p}\cdot{}^{p+1}_{p}\mathbf{p}){}^{p+1}_{p}\mathbf{l} - (2{}^{p+1}_{p}\mathbf{p}\cdot{}^{p+1}_{p}\mathbf{l}){}^{p+1}_{p}\mathbf{p})$ is of degree 1 or less in $\sin\theta_i, \cos\theta_i, i = 1,2,\ldots,p$. By our assumption that Property 4 is true for n equal to $p-1$, the entries of the matrix $\begin{pmatrix} \{({}^{p}_{1}\mathbf{p}\cdot{}^{p}_{1}\mathbf{p}){}^{p}_{1}\mathbf{l} - (2{}^{p}_{1}\mathbf{p}\cdot{}^{p}_{1}\mathbf{l}){}^{p}_{1}\mathbf{p}\}^T \\ \{({}^{p}_{1}\mathbf{p}\cdot{}^{p}_{1}\mathbf{p}){}^{p}_{1}\mathbf{m} - (2{}^{p}_{1}\mathbf{p}\cdot{}^{p}_{1}\mathbf{m}){}^{p}_{1}\mathbf{p}\}^T \\ \{({}^{p}_{1}\mathbf{p}\cdot{}^{p}_{1}\mathbf{p}){}^{p}_{1}\mathbf{n} - (2{}^{p}_{1}\mathbf{p}\cdot{}^{p}_{1}\mathbf{n}){}^{p}_{1}\mathbf{p}\}^T \end{pmatrix}^T$ are of degree 1 or less in $\sin\theta_i, \cos\theta_i, i = 1,2,\ldots,p-1$. The vector ${}^{p+1}_{p}\mathbf{l}$ is by definition of degree 1 or less in $\sin\theta_p, \cos\theta_p$. Therefore the term $\begin{pmatrix} ({}^{p}_{1}\mathbf{p}\cdot{}^{p}_{1}\mathbf{p}){}^{p}_{1}\mathbf{l} - (2{}^{p}_{1}\mathbf{p}\cdot{}^{p}_{1}\mathbf{l}){}^{p}_{1}\mathbf{p}^T \\ ({}^{p}_{1}\mathbf{p}\cdot{}^{p}_{1}\mathbf{p}){}^{p}_{1}\mathbf{m} - (2{}^{p}_{1}\mathbf{p}\cdot{}^{p}_{1}\mathbf{m}){}^{p}_{1}\mathbf{p}^T \\ ({}^{p}_{1}\mathbf{p}\cdot{}^{p}_{1}\mathbf{p}){}^{p}_{1}\mathbf{n} - (2{}^{p}_{1}\mathbf{p}\cdot{}^{p}_{1}\mathbf{n}){}^{p}_{1}\mathbf{p}^T \end{pmatrix}^T {}^{p+1}_{p}\mathbf{l}$ is of degree 1 or less in $\sin\theta_i, \cos\theta_i, i = 1,2,\ldots,p$. As a result, $({}^{p+1}_{1}\mathbf{p}\cdot{}^{p+1}_{1}\mathbf{p}){}^{p+1}_{1}\mathbf{l} - (2{}^{p+1}_{1}\mathbf{p}\cdot{}^{p+1}_{1}\mathbf{l}){}^{p+1}_{1}\mathbf{p}$ is of degree 1 or less in $\sin\theta_i, \cos\theta_i, i = 1,2,\ldots,p$. By similar arguments we may show that the other two functions $({}^{p+1}_{1}\mathbf{p}\cdot{}^{p+1}_{1}\mathbf{p}){}^{p+1}_{1}\mathbf{m} - (2{}^{p+1}_{1}\mathbf{p}\cdot{}^{p+1}_{1}\mathbf{m}){}^{p+1}_{1}\mathbf{p}$ and $({}^{p+1}_{1}\mathbf{p}\cdot{}^{p+1}_{1}\mathbf{p}){}^{p+1}_{1}\mathbf{n} - (2{}^{p+1}_{1}\mathbf{p}\cdot{}^{p+1}_{1}\mathbf{n}){}^{p+1}_{1}\mathbf{p}$ also exhibit this property. By induction on n, Property 4 is true for $n \geq 1$.

A2. Appendix 2.

A2.1 CLAIM. *The determinant of every* 6×6 *submatrix of* Σ'' *in (42) is an* 8th *degree polynomial in* x_3 *and not a rational function.*

A2.2 PROOF. The entries in the first four rows of the matrix Σ'' of (42) are quadratic polynomials in x_3 and the entries in the last two rows are rational functions of the form $\frac{quadratic\ polynomial\ in\ x_3}{(1+x_3^2)}$. Therefore, the determinant of any 6×6 submatrix of Σ'' is of the form $\frac{12^{th}\ degree\ polynomial\ in\ x_3}{(1+x_3^2)^2}$. We claim that the 12th degree polynomial in the numerator contains $(1+x_3^2)^2$ as a factor. If this is true then, the determinant is indeed an 8th degree polynomial. We proceed therefore to establish the claim that the above-mentioned 12th degree numerator polynomial does contain $(1+x_3^2)^2$ as a factor.

Let us examine the left-hand side of (39). The following table provides a summary of the basis elements of the left-hand sides of the fourteen scalar equations in (39).

Equation	Basis Elements
$\tilde{\mathbf{p}}$	$c_3f_1 + s_3f_2$, $s_3f_1 - c_3f_2$, f_3, 1
$\tilde{\mathbf{n}}$	$c_3r_1 + s_3r_2$, $s_3r_1 - c_3r_2$, r_3
$\tilde{\mathbf{p}} \cdot \tilde{\mathbf{p}}$	$\tilde{\mathbf{f}} \cdot \tilde{\mathbf{f}}$, $c_3f_1 + s_3f_2$, $s_3f_1 - c_3f_2$, f_3, 1
$\tilde{\mathbf{p}} \cdot \tilde{\mathbf{n}}$	$\tilde{\mathbf{f}} \cdot \tilde{\mathbf{r}}$, $c_3r_1 + s_3r_2$, $s_3r_1 - c_3r_2$, r_3
$\tilde{\mathbf{p}} \times \tilde{\mathbf{n}}$	$s_3(f_3r_1 - f_1r_3) + c_3(f_2r_3 - f_3r_2)$, $c_3(f_3r_1 - f_1r_3) - s_3(f_2r_3 - f_3r_2)$, $f_1r_2 - f_2r_1$, $s_3r_1 - c_3r_2$, $c_3r_1 + s_3r_2$, r_3
$(\tilde{\mathbf{p}} \cdot \tilde{\mathbf{p}})\tilde{\mathbf{n}} - (2\tilde{\mathbf{p}} \cdot \tilde{\mathbf{n}})\tilde{\mathbf{p}}$	$(ffr_1)c_3 + (ffr_2)s_3$, $(ffr_1)s_3 - (ffr_2)c_3$, (ffr_3), $s_3(f_3r_1 - f_1r_3) + c_3(f_2r_3 - f_3r_2)$, $c_3(f_3r_1 - f_1r_3) - s_3(f_2r_3 - f_3r_2)$, $f_1r_2 - f_2r_1$, $s_3r_1 - c_3r_2$, $c_3r_1 + s_3r_2$, r_3, $\tilde{\mathbf{f}} \cdot \tilde{\mathbf{r}}$, 1

In the above table the symbol (ffr_i) has been used as an abbreviation for the term $((\tilde{\mathbf{f}} \cdot \tilde{\mathbf{f}})\tilde{\mathbf{r}} - (2\tilde{\mathbf{f}} \cdot \tilde{\mathbf{r}})\tilde{\mathbf{f}})_i$. The first row of the table tells us that the left-hand side of each component equation of $\tilde{\mathbf{p}}$ is a linear combination of the terms $c_3f_1 + s_3f_2$, $s_3f_1 - c_3f_2$, f_3, 1. The remaining rows of the table provide similar information about $\tilde{\mathbf{n}}$, $\tilde{\mathbf{p}} \cdot \tilde{\mathbf{p}}$, $\tilde{\mathbf{p}} \cdot \tilde{\mathbf{n}}$, $\tilde{\mathbf{p}} \times \tilde{\mathbf{n}}$, $(\tilde{\mathbf{p}} \cdot \tilde{\mathbf{p}})\tilde{\mathbf{n}} - (2\tilde{\mathbf{p}} \cdot \tilde{\mathbf{n}})\tilde{\mathbf{p}}$. Since (40) is a linear combination of the scalar equations in (39), the left-hand side of (40) contains all the terms appearing in the right-hand column of the above table, viz. $c_3f_1 + s_3f_2$, $s_3f_1 - c_3f_2$, f_3, $c_3r_1 + s_3r_2$, $s_3r_1 - c_3r_2$, r_3, $\tilde{\mathbf{f}} \cdot \tilde{\mathbf{f}}$, $\tilde{\mathbf{f}} \cdot \tilde{\mathbf{r}}$, $s_3(f_3r_1 - f_1r_3) + c_3(f_2r_3 - f_3r_2)$, $c_3(f_3r_1 - f_1r_3) - s_3(f_2r_3 - f_3r_2)$, $f_1r_2 - f_2r_1$, $(ffr_1)c_3 + (ffr_2)s_3$, $(ffr_1)s_3 - (ffr_2)c_3$, (ffr_3), 1. We make the substitution, $s_3 = \frac{2x_3}{(1+x_3^2)}$, $c_3 = \frac{(1-x_3^2)}{(1+x_3^2)}$, (where $x_3 = tan(\frac{\theta_3}{2})$) in (40) and multiply each scalar equation by $(1+x_3^2)$ to clear denominators. We then make the substitution $x_3 = i$. Each scalar equation in (40) then becomes a linear combination of the four terms: f_1+if_2, r_1+ir_2, $(f_3r_1-f_1r_3)-i(f_2r_3-f_3r_2)$, $(ffr_1)+i(ffr_2)$. Since there is a total of six equations, clearly two of them must be linearly dependent on the remaining four. We may write these equations in matrix form as:

$$(\bar{\Sigma})\begin{pmatrix} s_4s_5 \\ s_4c_5 \\ c_4s_5 \\ c_4c_5 \\ s_4 \\ c_4 \\ s_5 \\ c_5 \\ 1 \end{pmatrix} = 0, \tag{68}$$

where $\bar{\Sigma}$ is a 4×9 array with row rank equal to four.

Next consider the matrix $\bar{\bar{\Sigma}}$ obtained from Σ'' of (42) by multiplying the last two rows by $(1 + x_3^2)$ in order to clear denominators. The row rank of $\bar{\bar{\Sigma}}$ is six because the six scalar equations in (42) are linearly independent. However, if we set $x_3 = i$ in $\bar{\bar{\Sigma}}$ then its row rank (and therefore column rank) becomes four, because $\bar{\bar{\Sigma}}_{|x_3=i}$ may be obtained from $\bar{\Sigma}$ by elementary column operations which cannot change the row rank. Therefore the determinant of every 6×6 submatrix of $\bar{\bar{\Sigma}}$ is a polynomial in x_3 which vanishes twice (rank deficiency $= 6 - 4 = 2$) at $x_3 = i$. By similar arguments, one may show that each such determinant also vanishes twice at $x_3 = -i$. Thus each 6×6 determinant of $\bar{\bar{\Sigma}}$ contains the factors $(x_3 + i)^2(x_3 - i)^2 = (1 + x_3^2)^2$. This proves our claim.

REFERENCES

ALBALA and ANGELES[1979]
H. Albala and J. Angeles, *Numerical solution to the input-output displacement equation of the general 7R spatial mechanism,* Proceedings of the Fifth World Congress on Theory of Machines and Mechanisms, pp. 1008-1011.

BUCHBERGER[1985]
B. Buchberger, *Gröbner Bases: An algorithmic method in polynomial ideal theory,* Chaper 6, in N.K. Bose (ed.), Multidimensional system theory, D. Reidel Publishing Co.

CANNY[1988]
J. Canny, *The complexity of robot motion planning,* MIT Press.

DENAVIT and HARTENBERG[1955]
J. Denavit and R. Hartenberg, *A kinematic notation for lower pair mechanisms based on matrices,* ASME Journal of Applied Mechanics, Vol. 77, pp. 215-221.

DUFFY and CRANE[1980]
J. Duffy and C. Crane, *A displacement analysis of the general spatial 7R mech-*

anism, Mechanisms and Machine Theory, Vol. 15, pp. 153-169.

FREUDENSTEIN[1973]
F. Freudenstein, *Kinematics: past, present and future,* Mechanisms and Machine Theory, Vol. 8, No. 2, pp. 151-161.

LEE and LIANG[1988a]
H. Lee and C. Liang, *A new vector theory for the analysis of spatial mechanisms,* Mechanisms and Machine Theory, Vol. 23, No. 3, pp. 209-217.

LEE and LIANG[1988b]
H. Lee and C. Liang, *Displacement analysis of the general spatial 7-link 7R mechanism,* Mechanisms and Machine Theory, Vol. 23, No. 3, pp. 219-226.

PAUL[1981]
R. Paul, *Robot manipulators: mathematics, programming, and control,* The MIT Press, Cambridge.

PIEPER[1968]
D. Pieper, *The kinematics of manipulators under computer control,* Ph.D. Thesis, Design Division, Stanford University.

RAGHAVAN and ROTH[1989]
M. Raghavan and B. Roth, *Kinematic analysis of the 6R manipulator of general geometry,* Proceedings of the Fifth International Symposium on Robotics Research (ISRR), Tokyo, August 1989.

ROTH, RASTEGAR and SCHEINMAN[1973]
B. Roth, J. Rastegar, and V. Scheinman, *On the design of computer controlled manipulators,* Proceedings of the First CISM-IFToMM Symposium, Vol. 1, pp. 93-113.

TSAI and MORGAN[1985]
L. Tsai and A. Morgan, *Solving the kinematics of the most general six- and five-degree-of-freedom manipulators by continuation methods,* Transactions of the ASME, Journal of Mechanisms, Transmissions, and Automation in Design, Vol. 107, pp. 189-200.

Power Systems Research Department, General Motors Research Laboratories, Warren, Michigan 48090

Proceedings of Symposia in Applied Mathematics
Volume **41**, 1990

Resolution of Kinematic Redundancy

J. Baillieul
D.P. Martin *

Abstract

Abstract: Kinematically redundant robotic mechanisms have more than the minimal number of degrees of freedom required to carry out a given class of tasks. This paper deals with kinematic redundancy in manipulator arms and describes a body of recent research on motion planning for such devices. A short history of the inverse kinematics problem for redundant manipulators begins with a discussion of the limitations which have recently been found with pseudo-inverse methods. Specifically, it is shown that such methods do not avoid kinematic singularities and are not "repeatable". The *extended Jacobian technique* is defined and discussed, and it is noted that this is essentially the only repeatable local method for joint space path generation. It is also noted that any method (such as the extended Jacobian technique) which resolves kinematic redundancy through kinematic constraints on joint space motions inevitably involves *algorithmic singularities*. The implications of such singularities for local planning methods are discussed. Computational methods for pathwise optimal resolution of kinematic redundancy are also presented. Necessary conditions for pathwise optimality are given in terms of boundary value problems, and it is shown by example that the true optimal solution must be selected from a family of mutually nonhomotopic locally optimal solutions. It is shown how homotopy continuation may be used both to simplify calculations which would otherwise require substantial computational resources and also to single out globally optimal joint space trajectories from among (infinitely) many locally optimal candidates satisfying the basic necessary conditions.

1 Introduction

This chapter provides a brief description of the current state of research on kinematically redundant robot arms. While considerations of time and space have forced us to be selective

1980 *Mathematics Subject Classification* (1985 *Revision*). Primary 34E20, 58C27, 70B15.
*The authors gratefully acknowledge the support of the U. S. Air Force through contracts and grants F33615-84-C-5131, AFOSR-85-0144, and AFOSR-90-0226.

in the emphasis given to various topics, we have tried to highlight a number of the most important contributions appearing in the last five years. The discussion is organized as follows: Section 1 describes the mathematical setting for the chapter and poses the principal mathematical problems involved in planning motions for kinematically redundant manipulators. At the end of the section, we discuss the important distinction between *local* and *global* methods for resolving kinematic redundancy. Section 2 highlights some past work on these problems and discusses recently discovered shortcomings in the once widely studied generalized inverse methods. Repeatable local methods for resolving kinematic redundancy involve functional constraints on joint variables, and in Section 3 we introduce the *extended Jacobian technique* for implementing and analyzing functionally constrained motions. In Section 4, we discuss global methods and observe that pathwise optimization leads to necessary conditions given in terms of boundary value problems. While solutions to such problems may involve a significant computational burden, we shall describe a computational approach based on *homotopy continuation* which recent experience has shown to work quite well on selected problems. A theoretical foundation for these methods is not complete at present, and the chapter concludes with results on the singularly perturbed necessary conditions which arise in defining a homotopy between a computationally complex global method (given in terms of a second order o.d.e.) and a less computationally demanding local method (where the necessary conditions may be described by purely algebraic equations).

Kinematic redundancy is encountered in robotics whenever a device or mechanism has more degrees of freedom than are necessary to accomplish a prescribed task. Thus kinematic redundancy will be a feature in multifingered robot hands, in manipulator arms mounted on mobile platforms, in systems involving multiple manipulator arms with shared work space, and in manipulators for which the number of joint degrees of freedom exceeds the minimal number of degrees of freedom required to perform a given task or class of tasks. Most of the issues and results of this paper pertain equally to all these instances of kinematic redundancy. The precise statement of results requires attention to the qualitative mechanical differences in the systems, however, and to avoid overly complex and heavily bifurcated statements of results, we shall only explicitly treat kinematic redundancy in manipulator arms. With little loss of generality, we shall also assume that the manipulators in question are single strand, open kinematic chains having *lower pair* joints. We recall that *lower pair* joints interconnect pairs of adjacent links allowing relative motions which are idealized as frictionless slipping along pairs of contact surfaces. Lower pair joints have played an important role historically in the theory and design of mechanisms, and recently the relative rigid body motions which they define have been characterized in terms of certain distinguished subgroups of $SE(3, \mathbf{R})$, the group of rigid motions of 3-space. Following [15], we make the following definition.

Definition 1.1 *A subgroup G of $SE(3, \mathbf{R})$ will be called a* joint subgroup *if there is a neighborhood U of the identity in $SE(3, \mathbf{R})$ and a pair of rigid bodies in contact such that inside U the set of all relative configurations of these bodies is identical to G.*

The six types of lower pair joints are then determined as follows.

Proposition 1.1 (Loncaric, [15]) *The only types of joint subgroups are $T(1)$ (=the set of all translations in a particular direction), $SO(2)$ (=the set of rotations about a given axis), $\bar{SO}(2)_p$ (=the group of screw motions with pitch p about a given axis), $SO(2) \times T(1)$, $SE(2, \mathbf{R})$, and $SO(3)$ (=the set of proper rotations of 3-space).*

In robot kinematics we study the relationship between joint configurations and *operational space* (=work space) configurations. For the open chain manipulator arms we study, operational space simply refers to the space of possible positions and orientations of an end plate, end effector, or tool. The joint configuration space, M, (or simply joint space) will be the product, $M = G_1 \times \cdots \times G_k$, of joint subgroups with the inherited differentiable structure. For actual manipulators, the joint space will typically be a proper subset of such a product group because of the inevitable joint limits which restrict the range of motions. The operational space, X, may also be modeled as a subset of a product of joint subgroups, and because we are considering open kinematic chain mechanisms, there will correspond to each joint configuration $\theta \in M$ a unique operational space configuration $x \in X$. We denote this functional relationship by writing

$$x = f(\theta). \tag{1}$$

At this level of abstraction, the inverse kinematics problem is to solve the equation (1) or more generally to find a right inverse for f. Because f is generally a many-to-one nonlinear function, solving (1) may be complicated and involve certain choices. The representation theory of $SE(3, \mathbf{R})$, however, supports explicit computations. A standard approach to representation of robot kinematics which implicitly exploits this theory has evolved from the work of Denavit and Hartenberg ([9]). More recent work of Brockett (This volume!), on the other hand, directly exploits the Lie group structure of $SE(3, \mathbf{R})$ to express kinematic relationships in terms of products of matrix exponentials. In either of these approaches, a coordinate frame is attached to each link in the kinematic chain. We may then represent the position and orientation of each of these coordinate frames with respect to its predecessor by means of a 4×4 ("homogeneous transformation") matrix, $\begin{pmatrix} Y & y \\ 0 & 1 \end{pmatrix}$. Here Y is a 3×3 orthogonal matrix prescribing the orientation and y is a 3-tuple giving the coordinates of the origin of the frame with respect to the preceding frame in the chain. Details of the Denavit-Hartenberg formalism may be found in many standard texts on robotics (e.g. [8]), and the elegant approach of Brockett is described in the first chapter of this book.

One rational approach to the inverse kinematics problem of associating a joint configuration θ to a given operational space configuration is to seek that value θ which optimizes some criterion function $g(\theta)$ subject to satisfying the constraint $f(\theta) = x$. Under reasonable assumptions, the mathematics underlying such pointwise optimization needn't involve more than multivariable calculus. If as is typical in robotic applications, however, there is given a path of configurations,

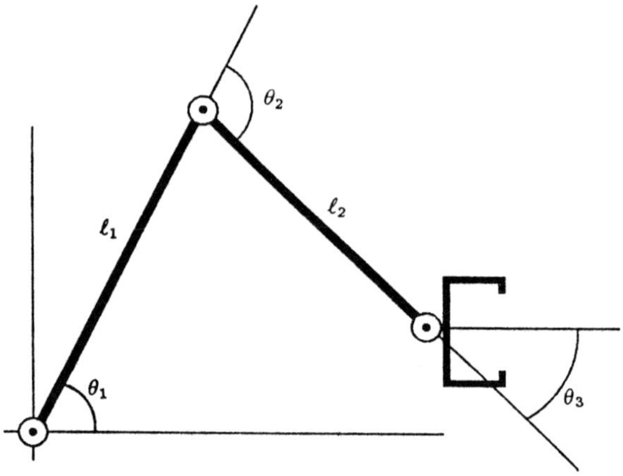

Figure 1: A planar manipulator with three revolute joints

$x(t)$, it is not always straightforward to generate a corresponding path $\theta(t)$, and the attendant mathematical issues will be discussed in the remainder of this paper.

1.1 Examples of Kinematic Redundancy and Mechanisms with Task Symmetries. Since the operational space X of any robotic device is a subset of the product of joint subgroups of $SE(3,\mathbf{R})$, it is convenient to discuss computational issues that arise from task symmetries in the forward and inverse kinematics problems in terms of group actions. If $X = G_1 \times \cdots \times G_k$ where each G_j is a subgroup of $SE(3,\mathbf{R})$, the natural *left action* $G \times X \to X$ is given by $g \circ (g_1, \ldots, g_k) = (gg_1, \ldots, gg_k)$, and the *right action* $X \times G \to X$ is given by $(g_1, \ldots, g_k) \circ g = (g_1 g, \ldots, g_k g)$. Let $G \subset SE(3,\mathbf{R})$ be a subgroup. We say there is a *(left) G-equivalence* between operational space configurations $x_1, x_2 \in X$ if there exists a $g \in G$ such that $x_2 = gx_1$. In making this definition, it is not stipulated that G be a joint subgroup, and as illustrated in the following examples, it is at times interesting to study G-equivalence relative to finite subgroups of $SE(3,\mathbf{R})$.

Example 1.1 We take the joint space for the planar manipulator depicted in Figure 1 to be $\mathbf{T}^3 = S^1 \times S^1 \times S^1$, where each joint coordinate $\theta_i \in S^1$ denotes the relative angle between corresponding successive links. The set of operational space configurations is the set of all possible positions and orientations of the end effector or hand with respect to some distinguished *base* coordinate frame. In Figure 1, the origin of the base frame coincides with the first joint, and the operational space may be thought of as

$$X = \left\{ \begin{pmatrix} x \\ y \\ \theta \end{pmatrix} : \begin{pmatrix} x \\ y \end{pmatrix} \in \mathbf{R}^2,\ 0 \le \theta < 2\pi,\ |\ell_1 - \ell_2| \le \sqrt{x^2 + y^2} \le \ell_1 + \ell_2 \right\}.$$

The functional relationship (1) between joint space and operational space takes the form

$$\begin{pmatrix} x \\ y \\ \theta \end{pmatrix} = \begin{pmatrix} \ell_1 \cos\theta_1 + \ell_2 \cos(\theta_1 + \theta_2) \\ \ell_1 \sin\theta_1 + \ell_2 \sin(\theta_1 + \theta_2) \\ \theta_1 + \theta_2 + \theta_3 \end{pmatrix}. \tag{2}$$

To emphasize the representation of X as a joint subgroup, we may also write the operational space X in terms of the standard 3×3 (homogeneous transformation) matrix representation of $SE(2, \mathbf{R})$:

$$X = \left\{ \begin{pmatrix} \cos\theta & -\sin\theta & x \\ \sin\theta & \cos\theta & y \\ 0 & 0 & 1 \end{pmatrix} : |\ell_1 - \ell_2| \le \sqrt{x^2 + y^2} \le \ell_1 + \ell_2,\ 0 \le \theta < 2\pi \right\}.$$

Apart from certain singular configurations (which we shall describe later), there are two distinct joint space configurations (elbow up; elbow down) corresponding via (2) to each operational space configuration.

Example 1.2 If we add a joint and link to the manipulator described in Example 1.1, we obtain a mechanism whose operational space remains a circular region in the plane and whose joint space is $M = \mathbf{T}^4 = S^1 \times S^1 \times S^1 \times S^1$. See Figure 5 below.) Note that $\dim M > \dim X$.

Because there are more joint space degrees of freedom than the minimal number required (=three in this case) to position and orient the manipulator's end effector arbitrarily in its operational space, the mechanism described in Example 1.2 is said to be *kinematically redundant.* The following examples will lead us to a somewhat more general definition of kinematic redundancy while at the same time pointing to the fact that many of the issues involved in motion planning for kinematically redundant mechanisms are also present in planning motions in the presence of task symmetries.

Example 1.3 Suppose the planar manipulator depicted in Figure 1 is given the task of inserting a $30° - 60°$ right triangular part into a hole with the same geometry. (See Figure 2.) The manipulator must match the part's position and orientation to that of the hole, and since there is only one way to do this, there are precisely two joint space configurations (elbow up and elbow down) allowing the part to be brought into coincidence with the hole.

Example 1.4 Suppose next the planar manipulator is given the task of inserting a rectangular part or work piece into a rectangular hole. (See Figure 3). Assume the center of the rectangle has coordinates $\begin{pmatrix} a \\ b \end{pmatrix}$ with respect to the base frame of the manipulator. Assuming the grasp

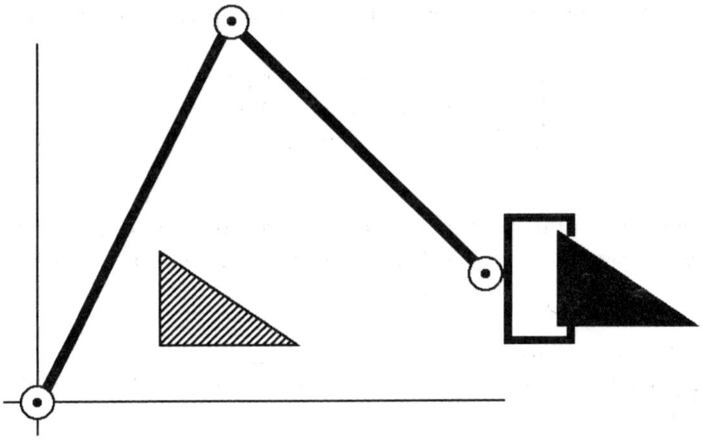

Figure 2: Planar manipulator inserting triangular part into a similarly shaped hole

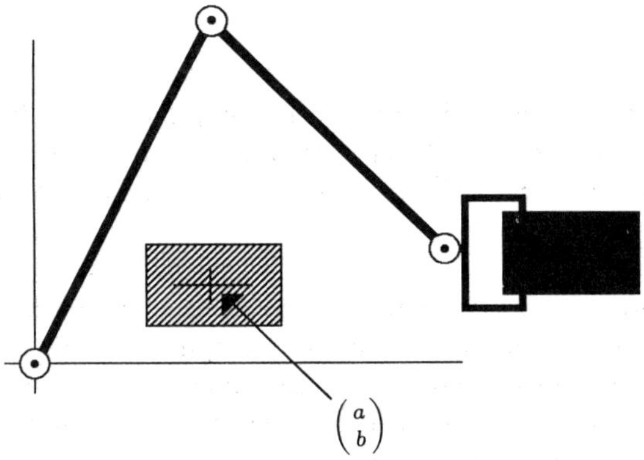

Figure 3: Planar manipulator inserting a rectangular part into a similarly shaped hole

depicted in the figure brings the part into coincidence with the hole when the end effector is positioned at

$$\begin{pmatrix} x \\ y \\ \theta \end{pmatrix},$$

it is not difficult to see that with the end effector is positioned and oriented at

$$\begin{pmatrix} \bar{x} \\ \bar{y} \\ \bar{\theta} \end{pmatrix} = \begin{pmatrix} 2a - x \\ 2b - y \\ \theta + \pi \end{pmatrix},$$

the part will also be coincident with the hole. There are thus four joint space configurations corresponding to the part being aligned with the hole.

Example 1.5 In the previous example, suppose the task is modified to insertion of a circular "peg" of radius r into a round hole also of radius r and centered at $\begin{pmatrix} a \\ b \end{pmatrix}$. The end effector is brought into coincidence with the hole for configurations of the end effector given by

$$\begin{pmatrix} x \\ y \\ \theta \end{pmatrix} = \begin{pmatrix} a - r\cos\phi \\ b - r\sin\phi \\ \phi \end{pmatrix}$$

for any ϕ.

The obvious symmetries in the two preceding examples may be formally described in terms of G-equivalence on operational space. This is most obvious when viewed from a local frame whose origin is located at $\begin{pmatrix} a \\ b \end{pmatrix}$ (=the center of the hole). With respect to this local frame, the task symmetry in Example 1.4 is described by a G-equivalence on operational space where G is a subgroup of $SE(2, \mathbf{R})$ which is isomorphic to $\mathbf{Z}_2$ and generated by the 180° rotation

$$\begin{pmatrix} -1 & 0 & 0 \\ 0 & -1 & 0 \\ 0 & 0 & 1 \end{pmatrix}.$$

The local frame G-equivalence of Example 1.5 is prescribed by

$$G = SO(2, \mathbf{R}) = \left\{ \begin{pmatrix} \cos\phi & -\sin\phi & 0 \\ \sin\phi & \cos\phi & 0 \\ 0 & 0 & 1 \end{pmatrix} : 0 \le \phi < 2\pi \right\}.$$

If operational space is described with respect to the base frame rather than the local frame, the symmetry groups of these examples are obtained by conjugation. For Example 1.4, G is generated by the element

$$\begin{pmatrix} 1 & 0 & a \\ 0 & 1 & b \\ 0 & 0 & 1 \end{pmatrix} \begin{pmatrix} -1 & 0 & 0 \\ 0 & -1 & 0 \\ 0 & 0 & 1 \end{pmatrix} \begin{pmatrix} 1 & 0 & -a \\ 0 & 1 & -b \\ 0 & 0 & 1 \end{pmatrix}$$

while for Example 1.5, G is the $SO(2)$ conjugate

$$\begin{pmatrix} 1 & 0 & a \\ 0 & 1 & b \\ 0 & 0 & 1 \end{pmatrix} \begin{pmatrix} \cos\phi & -\sin\phi & 0 \\ \sin\phi & \cos\phi & 0 \\ 0 & 0 & 1 \end{pmatrix} \begin{pmatrix} 1 & 0 & -a \\ 0 & 1 & -b \\ 0 & 0 & 1 \end{pmatrix}$$

where $0 \le \phi < 2\pi$. We note in both these examples that if

$$h = \begin{pmatrix} \cos\theta & -\sin\theta & x \\ \sin\theta & \cos\theta & y \\ 0 & 0 & 1 \end{pmatrix}$$

defines a position and orientation of the end effector for which the part is coincident with the hole, then any configuration gh with $g \in G$ also defines a part position/orientation which coincides with the hole.

Task symmetries of the type illustrated in these examples occur in many robotic applications. In welding for instance, any rotation of the welding torch about its longitudinal axis is irrelevant, and this task symmetry effectively reduces the dimension of the operational space. (Welding robots frequently in fact have five rather than six degrees of freedom precisely in order to take advantage of this task symmetry.) More generally, we find that when a task admits a G-equivalence, the problem of positioning and orienting the end effector in the operational space X may be simplified whenever there is no need to distinguish between G-equivalent configurations. (Distinctions among G-equivalent configurations will be important in applications where it is not possible to achieve all G-equivalent manipulator configurations—for instance, in tasks where a congested workspace obstructs certain configurations.) When such distinctions are not essential, we may simplify the mathematical analysis by formally identifying G-equivalent configurations. An important case arises when the orbits of G comprise a regular foliation of X. Except for certain singularities, X will in a wide variety of interesting cases be foliated by G-orbits. When this is the case, $\tilde{X} = X/G$ has a natural differentiable structure, and the kinematic mapping $f : M \rightarrow X$ determines a differentiable mapping $\tilde{f} = \pi \circ f : M \rightarrow \tilde{X}$ where $\pi : X \rightarrow \tilde{X}$ is the natural projection. $\tilde{X}$ is called the *reduced* operational space. Note that $\dim\tilde{X} \leq \dim X$, and the inequality is strict if $\dim G > 0$. We can now state the following formal definition.

Definition 1.2 *Let a robot manipulator be given with joint space M, (reduced) operational space X, and (reduced) kinematics mapping $f : M \rightarrow X$. If $\dim M > \dim X$ we say the manipulator is* kinematically redundant.

This definition enlarges the class of kinematically redundant manipulators to include mechanisms for which the redundancy is purely an artifact of symmetries in a designated task. When the task symmetry group has dimension 0, as it does in Example 1.4, the manipulator is not kinematically redundant in the formal sense. Nevertheless, the range of choices in solving the inverse kinematics problem is increased by any task symmetry, and as we shall note below, mechanisms with task symmetries present some of the same challenges for path planning as kinematically redundant manipulators.

1.2 Problems in Motion Planning. The principal problem treated in this paper involves finding a joint space path $\theta(\cdot)$ which corresponds via (1) to a given operational space path $x(\cdot)$. The simple examples which we have given illustrate the key feature common to planning joint space motion for both redundant and nonredundant manipulators: there are (possibly infinitely) many joint space paths corresponding to each operational space path. Criteria for selecting among these include:

- A joint trajectory $\theta(t)$ should describe a closed curve whenever the corresponding end effector trajectory $x(t)$ describes a closed curve.
- Joint trajectories should be as smooth as possible (In an actual robot, rough and jerky motions could damage the mechanism!), and the total motion of the joints should be no more than necessary to produce the prescribed end effector motion.
- Mechanical considerations limit the possible motions of any manipulator. Although we identify the motions of lower pair joints with joint subgroups, motions will typically be limited to some subset of the configurations specified by the group since the distance of travel in any joint will as a practical matter be limited, and of course the mechanism can never be configured such that solid members intersect one another. We thus adopt the rule of thumb that joint limits should be avoided by the widest possible margin in any joint space trajectory.
- Joint space configurations at which the Jacobian $J = \frac{\partial f}{\partial \theta}$ is rank deficient are called *kinematic singularities.* As we shall note below, when the manipulator configuration is close to a kinematic singularity, modest end effector velocities may unavoidably require very large joint velocities. Thus, we should try to specify joint trajectories which avoid singularities by a reasonable margin.
- Since there are choices to be made in specifying the motions of a kinematically redundant mechanism, these choices ought to optimize some rational objective. Implicit in the preceding three bullets are kinematic optimality criteria. Other criteria may be defined in terms of dynamic considerations (minimum acceleration, minimum required power from actuators, etc.).

The remark regarding kinematic singularities warrants some amplification. By differentiating the relationship (1), we obtain the velocity equation

$$\dot{x} = J\dot{\theta} \tag{3}$$

(where J is the Jacobian of f). If J is rank deficient (i.e. the mechanism is in a kinematically singular configuration), there are values of $\dot{x}$ for which no $\dot{\theta}$ satisfies (3). As a practical matter, one finds that near a kinematically singular configuration moderate values of $\|\dot{x}\|$ will require very large values of $\|\dot{\theta}\|$ in order to satisfy (3). For nonredundant manipulators, kinematic singularities are in general unavoidable. Example 1.1 illustrates this point since any configuration for which $\sin\theta_2 = 0$ is singular. Thus, if $\ell_1 = \ell_2$, bringing the hand into coincidence with the base forces the mechanism into a singular configuration. Example 1.2 is also of interest, since a straightforward calculation shows that any configuration such that $\sin\theta_2 = \sin\theta_3 = 0$ is singular. This illustrates a significant point which will be discussed further in the next section: adding more degrees of freedom to a mechanism does not eliminate the possibility of singular configurations.

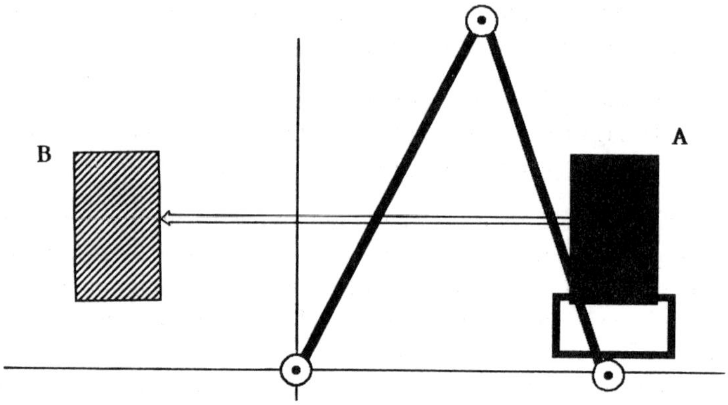

Figure 4: Manipulator task is to move the work piece from A to B along the indicated path

Whether or not a manipulator is kinematically redundant, we shall be guided by the objectives we have highlighted in choosing joint-space trajectories. In this regard, we shall examine two distinct approaches to resolving kinematic redundancy. *Local methods* solve (3) for $\dot{\theta}$ in terms of θ and x at each instant in time, and then generate an entire joint space path by integrating $\dot{\theta}$. *Global methods*, on the other hand, seek to define the path $\theta(\cdot)$ in terms of the entire operational space path $x(\cdot)$. Because global methods utilize the entire operational space path, they are necessarily noncausal and thus not well suited for incorporation into real time algorithms. Local methods, by contrast only use current information to determine the next step in planning a joint motion. In principle, they have the capacity to adapt to unforeseen changes in the operational space path, and they can be used in planning algorithms which utilize real time data from sensors. Moreover, as we shall note below, their implementation presents less daunting computational demands than global methods. The main disadvantage with local methods is the unavoidable occurrence of algorithmic singularities. (See Section 3.) These represent the ultimate performance penalty associated with an approach to planning which is in some reasonable sense optimal at each instant in time but which is incapable of incorporating information about future consequences of current motions. Local methods will be discussed in the next two sections, while global methods will be treated in Section 4.

Example 1.6 Consider the mechanism and work piece described in Example 1.4. Suppose we wish to move the work piece from position A to position B along the operational space path indicated in Figure 4 by grapsing it from above or below. If the work piece is grasped from below, there is considerably less overall motion required of the manipulator than if it is grasped from above because there is substantially less motion required from the first joint in the grasp-from-below motion. If the x-coordinates of A, B are respectively ϵ and $-\epsilon$, the relative difference in the motions required for grasp-from-below and grasp-from-above become more pronounced as $\epsilon \to 0$. Local methods are blind to such differences.

2 Kinematic Singularities and the Resolution of Kinematic Redundancy

The practical appeal of kinematically redundant robot manipulators is their enhanced potential for working in a congested environment and for being able to avoid operating near joint limits or kinematically singular configurations. Recall from the last section that if $\ell_1 = \ell_2$ in Example 1.1, then placing its end effector at the manipulator's base forces the manipulator into a singular configuration. The kinematically redundant manipulator of Example 1.2 was also seen to have kinematic singularities, but apart from those corresponding to the end effector being located on the boundary of the operational space, they are avoidable in the sense that whenever the mechanism is in a singular configuration, it may be reconfigured via a "self-motion"—a motion leaving the end effector unmoved—which moves the joints into a non-singular configuration. Two questions suggested by these example are: (i) is it ever possible to add an extra joint space degree of freedom so as to create a singularity-free robot mechanism, and (ii) is there any useful mathematical characterization to distinguish between *avoidable* and *unavoidable* singularities?

A partial answer to the second question has recently been given by Shamir ([22]) who uses the qualitative theory of differential equations to classify the singularities of 3-joint redundant manipulators. The first question has been answered in the case of manipulators with one degree of freedom revolute joints by Hollerbach ([11]) using a physical argument and by Gottlieb ([10]) using a topological argument. The following theorem was the first result on the question.

Theorem 2.1 (Brockett, [6]) *Any robot manipulator whose joints are of the $SO(2)$ (=one degree of freedom revolute) type will have kinematically singular configurations.*

Proof: In the context of planar mechanisms, our remarks regarding Example 1.2 can be extended in a straightforward manner to handle the general case. For spatial mechanisms, the formula

$$Y = R(\vec{k}_1, \theta_1)R(\vec{k}_2, \theta_2)\cdots R(\vec{k}_n, \theta_n) \tag{4}$$

gives the standard Brockett or Denavit-Hartenberg description of the orientation Y of the end effector in terms of the joint angles θ_j. Here $R(\vec{k}, \theta)$ designates the 3×3 proper orthogonal matrix corresponding to rotation θ in the counterclockwise sense about the axis defined by the unit vector $\vec{k}$.

We claim that for any three consecutive factors $R(\vec{k}_j, \theta_j)R(\vec{k}_{j+1}, \theta_{j+1})R(\vec{k}_{j+2}, \theta_{j+2})$ in this expression, there is a choice of θ_{j+1} such that $\vec{k}_j$ lies in the plane determined by $\vec{k}_{j+1}$ and $\vec{k}_{j+2}$. Suppose the coordinate frames have been chosen so that $\vec{k}_j$ is represented by $\begin{pmatrix} x_1 \\ x_2 \\ x_3 \end{pmatrix}$ in the

j-th coordinate frame in the chain, and $\vec{k}_{j+1}$ and $\vec{k}_{j+2}$ are represented respectively by $\begin{pmatrix} 0 \\ 0 \\ 1 \end{pmatrix}$ and $\begin{pmatrix} y_1 \\ y_2 \\ y_3 \end{pmatrix}$ in the $j+1$-st coordinate frame. Then $R(\vec{k}_{j+1}, \theta_{j+1})$ is represented by the matrix

$$\begin{pmatrix} \cos\theta_{j+1} & -\sin\theta_{j+1} & 0 \\ \sin\theta_{j+1} & \cos\theta_{j+1} & 0 \\ 0 & 0 & 1 \end{pmatrix}$$

with respect to the $j+1$-st frame, and the unit vector $\vec{k}_j$ is represented by

$$\begin{pmatrix} \bar{x}_1 \\ \bar{x}_2 \\ \bar{x}_3 \end{pmatrix} = \begin{pmatrix} \cos\theta_{j+1} & \sin\theta_{j+1} & 0 \\ -\sin\theta_{j+1} & \cos\theta_{j+1} & 0 \\ 0 & 0 & 1 \end{pmatrix} \begin{pmatrix} x_1 \\ x_2 \\ x_3 \end{pmatrix}$$

with respect to this frame. It is clear that either $\begin{pmatrix} x_1 \\ x_2 \\ x_3 \end{pmatrix} = \begin{pmatrix} 0 \\ 0 \\ 1 \end{pmatrix}$ or else there is some choice of θ_{j+1} such that $\begin{pmatrix} \bar{x}_1 \\ \bar{x}_2 \\ \bar{x}_3 \end{pmatrix}$ lies in the plane determined by $\begin{pmatrix} 0 \\ 0 \\ 1 \end{pmatrix}$ and $\begin{pmatrix} y_1 \\ y_2 \\ y_3 \end{pmatrix}$, in either case proving our claim.

Having established this claim, it then follows that for $j = 2, \ldots, n-1$, joint angles θ_j may be chosen such that all joint axes in the mechanism lie in the same plane. This is equivalent to saying that all columns in the $3 \times n$ Jacobian matrix associated with the mapping determined by the equation (4) will be orthogonal to a given vector (normal to the plane in question). Hence the Jacobian will be rank deficient, proving the theorem.□

2.1 Generalized Inverse Methods for Resolving Kinematic Redundancy. An early approach to resolution of kinematic redundancy in robot manipulators following a prescribed operational space path $\dot{x}(t)$ was to solve (3) for the joint velocity $\dot{\theta}$ of minimum norm $\|\dot{\theta}\|$. This idea, when it was proposed by Whitney in 1969 ([23]), appeared to have two attractive features. First it seemed reasonable to hope that by minimizing $\|\dot{\theta}\|$ at each point in time, kinematic singularities would be avoided automatically, and second, because we can write the norm-minimizing solution explicitly as $\dot{\theta} = J^{\dagger}\dot{x}$, where $J^{\dagger} = J^T(JJ^T)^{-1}$ is the Moore-Penrose generalized inverse of the Jacobian, planning algorithms could be implemented recursively using standard numerical methods for ordinary differential equations. Unfortunately, closer analysis of the Moore-Penrose generalized inverse approach to motion planning for kinematically redundant manipulators has revealed several initially unanticipated problems. The following theorem shows that the approach does not generate joint-space trajectories which avoid kinematic singularities in any practical sense.

Theorem 2.2 (Baillieul) *Consider a manipulator whose joint space and operational space configurations are related by (1). Let* $\epsilon > 0$ *be given. Let* θ_s *be a point at which* $J = J(\theta)$ *is rank deficient (i.e.* θ_s *defines a kinematically singular configuration), and let* θ^* *be a point such that* $\|\theta^* - \theta_s\| < \epsilon$*. Finally, let* $x^* = f(\theta^*)$*. Then for any operational space configuration* x_0 *and any* C^1 *curve* $x(\cdot)$ *initiating at* x_0 *and terminating at* x^**, there is a curve* $\theta(\cdot)$ *corresponding to* $x(\cdot)$ *via*

$$\dot{\theta} = J^{\dagger}\dot{x} \tag{5}$$

which terminates at θ^**.*

Proof: Suppose ϵ, θ_s, θ^*, x_0, x^*, and the curve $x(\cdot)$ satisfy the hypothesis of this theorem. By reparameterization if necessary, we may assume $x(0) = x_0$ and $x(1) = x^*$. Define a reverse trajectory in operational space by $y(t) = x(1-t)$ for $0 \le t \le 1$, and define a trajectory $\bar{\theta}(\cdot)$ in joint space by means of the differential equation

$$\dot{\bar{\theta}} = J^{\dagger}\dot{y}, \qquad \bar{\theta}(0) = \theta^*.$$

Then one easily verifies that the curve $\theta(t) = \bar{\theta}(1-t)$ satisfies the conclusion of the theorem.□

Another practical problem with the Moore-Penrose generalized inverse approach has been pointed out by Klein and Huang ([14]), whose simulations showed that closed trajectories in operational space did not generally give rise to closed joint space trajectories. This feature is a matter of practical concern since there is uncertainty about how the robot joints will configure themselves after several repeated executions of a task cycle. Topological issues involved in associating closed joint space paths to closed operational space paths have been treated in [5]. In [21] it has been shown that a nonintegrable relationship between operational space paths and joint space paths can be interpreted in terms of the non-involutivity of a certain distribution of vector fields. This repeatability problem was solved in 1985 ([1]) by the *extended Jacobian method*, details of which will be given in the next section.

2.2 Pointwise Optimization Criteria and Functional Constraints for Resolving Kinematic Redundancy. We conclude this section by discussing the role optimization has played in various approaches to resolving kinematic redundancy. A great many kinematic optimality criteria have been proposed in the literature, including minimizing joint space velocity, avoiding obstacles by the greatest possible amount with respect to various distance measures, avoiding joint limits, and avoiding kinematic singularities. (See, for example, [20].) Kinematically singular configurations are determined by the vanishing of $\det(JJ^T)$, and criteria for avoiding singularities by the widest possible margin include minimizing $g_1(\theta) = tr((JJ^T)^{-1})$ (Baillieul, [2]), maximizing $g_2(\theta) = \lambda_{\min}(JJ^T)$ (=smallest eigenvalue of JJ^T, Klein, [13]), and maximizing $g_3(\theta) = \sqrt{\det(JJ^T)}$, the *manipulatability index* (Yoshikawa, [25]). If we are given a kinematic function f, then according to the following theorem any optimality criterion g of this form defines an "optimality surface" which effectively resolves the redundancy.

Theorem 2.3 (Baillieul, [1]) *A necessary condition for θ_0 to define a local maximum of an objective function $g(\theta)$ subject to the positional constraint $f(\theta) = x$ is that*

$$G(\theta_0) = L_{\mathbf{n}} g\Big|_{\theta_0} = 0$$

for all vectors $\mathbf{n}$ in the null space of $J(\theta)$ where $L_{\mathbf{n}} g$ denotes the Lie derivative of the function g in the direction of the vector $\mathbf{n}$.

Proof: Suppose that θ_0 is a local maximum of $g(\theta)$ where θ is restricted to lie in the fiber $\{\theta : f(\theta) = x\}$ for fixed x. We call curves in this fiber "self-motions," and note that to first order, self-motions starting at θ_0 are in the direction of the null space of $J(\theta_0)$. The conclusion of the theorem follows from the observation that if θ_0 defines a local maximum in the fiber, the first order change in g along any self-motion through θ_0 must be zero.□

The importance of optimality criteria for functionally constraining the motions of redundant mechanisms (and thus resolving the redundancy) is not completely revealed by this theorem. The use of constraints which do not arise in conjunction with some form of optimization can inhibit motions of the mechanism in such a way as to result in significant effective reduction of work space volume. The reader may wish to construct examples of this phenomenon.

For a variety of reasons, it is desirable to understand how to define optimality criteria in a coordinate-free way. The need for coordinate-free definitions is well appreciated by geometers and may be understood quite simply in the present context as follows. Suppose the position of the end effector of a manipulator is prescribed in terms of its joint angle settings by

$$x = f(\theta)$$

where x is an m-vector, and θ is an $n > m$-vector. The redundancy may be resolved by constraining the joint variables to an m-dimensional surface. Suppose such a surface has a local parameterization of the form $\theta = \theta(u)$, where $u \in \mathbf{R}^m$. Then locally the above positional relationship is given by

$$x = G(u)$$

where $G(u) = f(\theta(u))$. It is thus tempting to define, say, the constrained analog of the above manipulatability index $g_3(\theta)$ by writing

$$g(u) = \left(\det\left[(J_u G)(J_u G)^T\right]\right)^{\frac{1}{2}}$$

where $J_u G = \frac{\partial G}{\partial u}$, but this is easily seen to depend on the chosen parameterization $\theta = \theta(u)$. It does not provide a meaningful measure of the degree to which the motion is restricted by constraining the joint variables to lie on the surface. The following theorem will be useful in removing this undesirable coordinate dependence.

Theorem 2.4 *The criteria given above are equivalent to the following:*

$$g_1(\theta) = \int_{S^{m-1}} \|J(\theta)^+ v\|^2 \, dv$$

$$g_2(\theta) = \frac{1}{\max_{v \in S^{m-1}} \|J(\theta)^+ v\|^2}$$

$$g_3(\theta) = \lim_{\epsilon \to 0} \sum_k \frac{(vol_m\, f(F_k(\theta,\epsilon)))^2}{(vol_m\, F_k(\theta,\epsilon))^2}$$

where $F_k(\theta,\epsilon)$ is the k-th m-dimensional slice (parallel to the k-th coordinate m-plane) through a standard n-cube with linear dimension ϵ, centered at θ, and vol_m denotes m-dimensional volume. The sum here is taken over the $\binom{n}{m}$ standard coordinate m-planes in $\mathbf{R}^n$. S^{m-1} denotes the unit m-1-sphere in $\mathbf{R}^m$.

The reader is encouraged to work out a proof of this theorem for herself. The case of g_3 is proved in [2]. An important observation made in [2] is that $\det(JJ^T)$ = *sum of squares of $m \times m$ minors of J.* I.e.

$$\det(JJ^T) = \sum_{i_1 < \dots < i_m} \left|\frac{\partial(f_1 \dots f_m)}{\partial(\theta_{i_1} \dots \theta_{i_m})}\right|^2. \tag{6}$$

Each term in the sum is the square of the Jacobian determinant of a certain mapping. Thus, as suggested by the theorem, it may be thought of as a certain ratio of m-dimensional volumes swept out in operational space to corresponding volumes in the joint space. This point of view turns out to be crucial in extending the definition of the manipulatability index to situations where the manipulator's joint configurations are constrained to a hypersurface $G(\theta) = 0$. To make this extension, we must be able to compute the manipulatability index relative to surface coordinates.

Suppose that in a neighborhood of some point on the surface we may solve $G(\theta) = 0$ to write $\theta_n = \theta_n(\theta_1, \dots, \theta_{n-1})$. In this neighborhood, the kinematics function f may be thought of as a function of $n-1$ variables: $\hat{f}(\theta_1, \dots, \theta_{n-1}) = f(\theta_1, \dots, \theta_{n-1}, \theta_n(\theta_1, \dots, \theta_{n-1}))$. Each term in the sum (6) is not replaced by $|\frac{\partial(\hat{f}_1 \dots \hat{f}_m)}{\partial(\theta_{i_1} \dots \theta_{i_m})}|^2$ but rather by $|\frac{\partial(\hat{f}_1 \dots \hat{f}_m)}{\partial(\theta_{i_1} \dots \theta_{i_m})}|^2 / V_{i_1 \cdots i_m}$ where $V_{i_1 \cdots i_m}$ is the square of the element of m-dimensional volume in the surface: $V_{i_1 \cdots i_m} = (1 + (\frac{\partial \theta_n}{\partial \theta_{i_1}})^2 + \dots + (\frac{\partial \theta_n}{\partial \theta_{i_m}})^2)$. Now one may show that

$$\left|\frac{\partial(\hat{f}_1 \dots \hat{f}_m)}{\partial(\theta_{i_1} \dots \theta_{i_m})}\right| = \det\begin{pmatrix} \frac{\partial f}{\partial \theta_{i_1}} & \dots & \frac{\partial f}{\partial \theta_{i_m}} & \frac{\partial f}{\partial \theta_n} \\ \frac{\partial G}{\partial \theta_{i_1}} & \dots & \frac{\partial G}{\partial \theta_{i_m}} & \frac{\partial G}{\partial \theta_n} \end{pmatrix} / \frac{\partial G}{\partial \theta_n}.$$

Hence

$$\left|\frac{\partial(\hat{f}_1 \dots \hat{f}_m)}{\partial(\theta_{i_1} \dots \theta_{i_m})}\right|^2 / V_{i_1 \cdots i_m} = \frac{\left(\det\begin{pmatrix} \frac{\partial f}{\partial \theta_{i_1}} & \dots & \frac{\partial f}{\partial \theta_{i_m}} & \frac{\partial f}{\partial \theta_n} \\ \frac{\partial G}{\partial \theta_{i_1}} & \dots & \frac{\partial G}{\partial \theta_{i_m}} & \frac{\partial G}{\partial \theta_n} \end{pmatrix}\right)^2}{\frac{\partial G}{\partial \theta_{i_1}}^2 + \dots + \frac{\partial G}{\partial \theta_{i_m}}^2 + \frac{\partial G}{\partial \theta_n}^2}.$$

Although this calculation suggests that we might define the constrained manipulatability index of the kinematics function f relative to the constraint function G by summing terms of this type, such an approach would clearly reflect the distinguished role the coordinate θ_n played in the above calculation. In generic situations, however, the above calculation may be repeated repeated for each coordinate θ_k, and the results may be averaged to produce the desired invariant quantity. This motivates the following definition.

Definition 2.1 *Consider a kinematically redundant manipulator having kinematics function f with joint space dimension n and reduced operational space dimension m. The* constrained manipulatability index *associated to f and the constraint $G(\theta) = 0$ is given by $m_G(\theta) =$*

$$\left(\sum_{i_1 < i_2 < \cdots < i_{m+1}} \frac{\left(\det \begin{pmatrix} \frac{\partial f}{\partial \theta_{i_1}} & \cdots & \frac{\partial f}{\partial \theta_{i_{m+1}}} \\ \frac{\partial G}{\partial \theta_{i_1}} & \cdots & \frac{\partial G}{\partial \theta_{i_{m+1}}} \end{pmatrix} \right)^2}{\frac{\partial G}{\partial \theta_{i_1}}^2 + \cdots + \frac{\partial G}{\partial \theta_{i_{m+1}}}^2} \right)^{\frac{1}{2}}.$$

A detailed exploration of the practical use of this index will not be pursued (see [2]), except that we mention our primary motivation in defining this quantity has been to study constraint hierarchies. In the following section, it will be noted that *algorithmic singularities* associated with the extended Jacobian technique lie on the zero locus of the constrained manipulatability index.

Remark 2.1 In the special case $n = m + 1$, the constrained manipulatability index is written as

$$m_G(\theta) = \frac{\left| \det \begin{pmatrix} \frac{\partial f}{\partial \theta_1} & \cdots & \frac{\partial f}{\partial \theta_n} \\ \frac{\partial G}{\partial \theta_1} & \cdots & \frac{\partial G}{\partial \theta_n} \end{pmatrix} \right|}{\|\frac{\partial G}{\partial \theta}\|}. \tag{7}$$

In this case, the constrained manipulatability index may be interpreted in terms of the Riemannian metric induced on the surface $G(\theta) = 0$ by the natural Riemannian metric on the ambient unconstrained joint space. Note that for each θ such that $G(\theta) = 0$,

$$\mathcal{G} = I - \frac{G_\theta G_\theta^T}{\|G_\theta\|^2}$$

is the orthogonal projection onto the tangent space of $G(\theta) = 0$ at θ. (Here G_θ denotes the column vector of partial derivatives $\frac{\partial G}{\partial \theta_j}$, $j = 1, \ldots, n$, evaluated at θ.) $\mathcal{G}$ is symmetric, and if $\theta(t)$ defines a differentiable curve on $G(\theta) = 0$, then $\|\dot\theta(t)\|^2 = \dot\theta(t)^T \mathcal{G} \dot\theta(t)$. Hence the $n \times n$-matrix $\mathcal{G}$ defines the inherited Riemannian metric tensor with respect to θ-coordinates on the surface $G(\theta) = 0$. A simple argument using the properties of determinants shows that $m_G(\theta)$ in (7) may be rewritten as

$$m_G(\theta) = \sqrt{\det(J\mathcal{G}J^T)}.$$

This may be thought of as the standard manipulatability index of $\hat{J} = \mathcal{G}J$, the Jacobian projected onto the tangent space of $G(\theta) = 0$. Note, however, that when $n > m + 1$ our more general definition of $m_G(\cdot)$ does not have such a simple interpretation.

In addition to kinematic optimality criteria of the type we have discussed, a number of authors have pursued optimization of dynamic and kinetic quantities to resolve the motion of kinematically redundant manipulators. Hollerbach and Suh ([12]) for instance, have studied the motions of trajectories designed to keep torque demands on all joint actuators within prescribed ranges. Unfortunately, simulations ([12]) have shown that such torque-based resolution of redundancy may produce unforeseen and dramatically unstable motions. The problem, of course, is that all locally optimal methods for resolving redundancy ignore the future, and by optimizing a criterion at each instant, they may in fact prescribe motions which are disastrous from a global point of view. We have seen an example of local optimization resulting in poor global characteristics above where we showed that minimizing the velocity at each instant would not generate joint space trajectories which avoided kinematic singularities. Maciejewski ([16]) has discussed similar pathologies arising in motions based on local torque minimization.

3 Local Methods for Resolution of Kinematic Redundancy—the *Extended Jacobian Technique*

Local methods for generating joint space trajectories $\theta(\cdot)$ corresponding to given operational space trajectories $x(\cdot)$ have the dual advantages of being easy to implement and having the capacity to adapt to unforeseen changes in $x(\cdot)$. In this section we shall introduce the *extended Jacobian technique* which we originally proposed in [1]. This method meets the requirement of being repeatable, and it naturally incorporates kinematic optimality criteria of the type introduced in the last section. The recent work of Shamir and Yomdin ([21]) shows that this is essentially the only repeatable local method.

To incorporate an objective function g which is to be optimized in carrying out the motion, note that Theorem 2.3 implies there will be $n - m$ constraints $G_j(\theta) = 0$, $j = 1, \ldots, n - m$, where $G_j(\theta) = L_{n_j}g$, the Lie derivative of g in the j-th independent direction in the Jacobian null space. If the end effector traces a trajectory $x(t)$ along which the corresponding joint configuration $\theta(t)$ extremizes g at each point, then we have

$$\begin{pmatrix} f(\theta) \\ G_1(\theta) \\ \vdots \\ G_{n-m}(\theta) \end{pmatrix} = \begin{pmatrix} x(t) \\ 0 \\ \vdots \\ 0 \end{pmatrix}.$$

Differentiating both sides provides a (forward) kinematic relationship between joint angle velocities and end effector velocity:

$$\begin{pmatrix} \frac{\partial f}{\partial \theta} \\ \cdots \\ \frac{\partial G}{\partial \theta} \end{pmatrix} \cdot \dot{\theta}(t) = \begin{pmatrix} \dot{x}(t) \\ 0 \end{pmatrix}. \tag{8}$$

The coefficient matrix

$$J_e = \begin{pmatrix} \frac{\partial f}{\partial \theta} \\ \cdots \\ \frac{\partial G}{\partial \theta} \end{pmatrix}$$

is square, and we shall call this the *extended Jacobian.*

Provided the extended Jacobian is nonsingular along a trajectory of interest, we may solve the inverse kinematics problem by writing

$$\dot{\theta}(t) = J_e^{-1} \begin{pmatrix} \dot{x}(t) \\ 0 \end{pmatrix}.$$

Obviously the extended Jacobian will be singular at any point where $J = \frac{\partial f}{\partial \theta}$ is singular. Other singular points of the extended Jacobian are characterized by the following.

Theorem 3.1 (Baillieul, [3]) *Suppose θ_0 is a joint space configuration at which J has full rank m. Then J_e is singular if and only if the linear operator defined by the $(n-m) \times n$ matrix $\frac{\partial G}{\partial \theta}$ restricted to the null space of J has rank $< n-m$.*

Proof: Suppose J_e is singular. If $\frac{\partial G}{\partial \theta}(\theta_0)$ has rank $< n-m$, it has rank $< n-m$ *a fortiori* when restricted to the subspace $\ker J$. If $J_e(\theta)$ is singular but both J and $\frac{\partial G}{\partial \theta}$ have full rank, there is a row vector $v \neq 0$ such that $v \cdot J_e(\theta_0) = 0$, and such that v has nonzero components v_i with $i \leq m$ and also nonzero components v_i with $i > m$. But this implies there is an $(n-m)$-tuple $\bar{v} \neq 0$ such that $\bar{v} \cdot \frac{\partial G}{\partial \theta}(\theta_0)$ lies in the row space of $J(\theta_0)$. Then $\bar{v} \cdot \frac{\partial G}{\partial \theta}(\theta_0) \cdot \hat{n} = 0$ for all $\hat{n} \in \ker J(\theta_0)$, and hence $\frac{\partial G}{\partial \theta}(\theta_0)$ has rank $< n-m$ on $\ker J(\theta_0)$.

Suppose, on the other hand, that $\frac{\partial G}{\partial \theta}(\theta_0)$ has rank $< n-m$ on $\ker J(\theta_0)$. Let N be an $(n-m) \times n$ matrix whose rows span $\ker J(\theta_0)$. We may write

$$J_e \cdot (J^{\dagger} \vdots N^T) = \begin{pmatrix} I & 0 \\ A(\theta) & B(\theta) \end{pmatrix}, \tag{9}$$

and since $B(\theta_0) = \frac{\partial G}{\partial \theta}(\theta_0)N^T$ has rank $< n - m$ and *rank* $J_e = m +$ *rank* $B(\theta_0)$, it follows $J_e(\theta_0)$ is singular, proving the theorem.□

It was shown in Baillieul *et al.* ([4]) that any C^1 joint space trajectory not passing through a singularity of J could be generated by an appropriate choice of $v(\cdot)$ in

$$\dot{\theta} = J^{\dagger}\dot{x} + N^T v(t). \tag{10}$$

The problem of specifying $v(\cdot)$ in some useful way can be approached via the *extended Jacobian technique*. Multiplying both sides of (9) by the composite vector $\begin{pmatrix} \dot{x} \\ v \end{pmatrix}$, we see from (9) and (10) that

$$0 = A(\theta)\dot{x} + B(\theta)v. \tag{11}$$

If J_e is invertible, we have seen that $B(\theta)$ has rank $n - m$. In this case, v can always be found. If J_e is not invertible, it may still be possible to find a v which solves (11), but the conditions are more delicate. This motivates the following.

Definition 3.1 *Points in the joint space where (11) cannot be solved for some value of $\dot{x}$ are called* algorithmic singularities *associated with the constraint(s)* $G_1(\theta) = 0, \dots, G_{n-m}(\theta) = 0$.

It is important to emphasize that algorithmic singularities are *not* artifacts of the extended Jacobian technique. Unfortunately, they are inevitably introduced if functional constraints (or optimality criteria) are used to resolve the redundancy in almost any manipulator having revolute joints.

Example 3.1 Suppose the optimality criterion is the *manipulatability index* $g(\theta) = \sqrt{\det(JJ^T)}$ discussed in the previous section. Consider the (redundant) planar manipulator of Example 1.2. The position of the end effector with respect to the base is given by

$$\begin{pmatrix} x \\ y \end{pmatrix} = \begin{pmatrix} \ell_1 \cos\theta_1 + \ell_2 \cos(\theta_1 + \theta_2) + \ell_3 \cos(\theta_1 + \theta_2 + \theta_3) \\ \ell_1 \sin\theta_1 + \ell_2 \sin(\theta_1 + \theta_2) + \ell_3 \sin(\theta_1 + \theta_2 + \theta_3) \end{pmatrix}.$$

Maximizing the manipulatability index for this kinematics function is equivalent to maximizing $g(\theta) = \sin^2\theta_2 + \sin^2\theta_3$. This defines the constraint $G(\theta) = L_g\hat{n} = 0$ in terms of which the extended Jacobian is defined. (When $\ell_j = 1$ for $j = 1, 2, 3$, this constraint is simply $\theta_2 - \theta_3 = 0$.) As noted in [1], the algorithmic singularities for this criterion occur when the end effector touches the base, forming a closed kinematic chain allowing no motions in joint variables θ_2 and θ_3. These algorithmic singularities may also be interpreted in terms of the constrained manipulatability index, $m_G(\theta) = |\det(J_e)|/\|\frac{\partial G}{\partial \theta}\|$, of Section 2. If $\ell_j = 1$ for $j = 1, 2, 3$, we have $m_G(\theta) = \sqrt{2} \cdot |\sin\theta_2| \cdot |1 + 2\cos\theta_2|$. Setting this equal to zero, we find the singularities associated with the criterion occur when $\sin\theta_2 = 0$ (which are the singularities of J) and when $\theta_2 = \pm 2\pi/3$ which are the algorithmic singularities associated with the constraint.

4 Homotopy Continuation Methods for Pathwise Resolution of Kinematic Redundancy

Recently, attention has turned to global optimization as a means of resolving kinematic redundancy. Nakamura and Hanafusa ([19]) describe this approach and present an algorithm for computing globally optimal joint space trajectories for redundant manipulators. In the present section, we present an alternative formulation of the necessary conditions, describe qualitative features of extremal trajectories, and discuss the theoretical foundations of *homotopy continuation* as an approach to algorithm design for these computationally rigorous problems.

We begin with the velocity equation (3). Our approach is to formulate the problem of finding a joint-space trajectory as a constrained optimization problem where we choose to minimize the path integral

$$\int_0^T (\frac{1}{2}\dot{\theta}^T W^{-1}\dot{\theta} + g(\theta))\, dt \tag{12}$$

subject to the kinematic constraint $x(t) = f(\theta(t))$. In this criterion, we assume the weighting matrix W is symmetric, and it is allowed to have an explicit dependence on t.

Theorem 4.1 *Joint space trajectories which optimize (12) satisfy*

$$\ddot{\theta} = J_W^{\dagger}(\ddot{x} - \dot{J}\dot{\theta}) + \mathbf{P}_W(\dot{W}W^{-1}\dot{\theta} + Wg_{\theta}) \tag{13}$$

where

$$\mathbf{P}_W = (I - J_W^{\dagger}J)$$

is the weighted null space projection operator of the Jacobian, J, and

$$J_W^{\dagger} = WJ^T(JWJ^T)^{-1}$$

is the weighted pseudo-inverse of J.

Proof: This theorem follows from a variational argument in which the joint space trajectory variations $\delta\theta$ are constrained to lie in the direction of the null space of J. The Euler-Lagrange operator defined by the functional (12) is

$$\frac{d}{dt}(W^{-1}\dot{\theta}) - \frac{\partial g}{\partial \theta},$$

and if $\theta(\cdot)$ minimizes (12) over all trajectories which vary with respect to $\theta(\cdot)$ in the direction $\ker J(\theta(t))$—i.e. with respect to all trajectories corresponding to the given operational space path $x(\cdot)$—, it follows that $\theta(\cdot)$ must satisfy

$$\mathbf{P}_W\{W[\frac{d}{dt}(W^{-1}\dot{\theta}) - \frac{\partial g}{\partial \theta}]\} = 0. \tag{14}$$

Note that $\mathbf{P}_W$ is the orthogonal projection relative to the inner product defined by the symmetric positive definite matrix W^{-1}. If we differentiate (1) twice with respect to t, we obtain

$$\ddot{x} = \dot{J}\dot{\theta} + J\ddot{\theta}$$

which may be equivalently written as

$$\ddot{\theta} = J_W^{\dagger}(\ddot{x} - \dot{J}\dot{\theta}) + \mathbf{P}_W \nu \tag{15}$$

for an appropriate choice of ν. Multiplying both sides of this equation by the projection operator $\mathbf{P}_W$ and noting that $\mathbf{P}_W J_W^{\dagger} = 0$, we obtain

$$\mathbf{P}_W \ddot{\theta} = \mathbf{P}_W \nu.$$

Using (14), we find that this implies

$$\mathbf{P}_W \nu = \mathbf{P}_W (\dot{W} W^{-1} \dot{\theta} + W \frac{\partial g}{\partial \theta}),$$

and substituting this into (15) proves the theorem.□

In order to uniquely specify the optimal solution, one must consider boundary conditions as well as the necessary condition provided by (13). When we first began studying integral criteria for resolving kinematic redundancy, it had been hoped that by minimizing the integral criterion (12) we would find joint-space trajectories that met the kinematic constraint (1) while staying more than a minimum distance from kinematic singularities. Unfortunately, as has been noted in [18], solutions to (13) may include singular points if the boundary conditions are chosen in a sufficiently malicious way. In what follows, we shall treat the case of periodic boundary conditions. This is natural since we are treating periodic operational space trajectories. Thus if x satisfies $x(0) = x(T)$, the objective is to find trajectories which statisfy (13) subject to $\theta(0) = \theta(T)$ and $\dot{\theta}(0) = \dot{\theta}(T)$.

Example 4.1 The Existence of Multiple Nonhomotopic Extremal Solutions. Pathwise optimization with periodic boundary conditions may be applied to path planning for the manipulator studied in Example 1.2. This geometry has been chosen for the purpose of illustration since it is especially simple to observe qualitative differences in comparing optimal and non-optimal trajectories. The necessary conditions expressed in terms of the system of differential equations (13) together with periodic boundary conditions will, in general, be sufficient to determine a family of trajectories that extremize the path integral (12). In the specific example treated here, we have chosen $g(\theta) \equiv 0$ and $W(t) \equiv I$. Figure 5 shows the geometry of the planar three bar manipulator, and the trajectory in the operational space that it is to trace. The endpoint of the manipulator is to describe the indicated circle in unit time, in a counterclockwise direction. Figures 6 and 7 show the joint angle time histories of two different trajectories that satisfy equation (13) and are periodic. The trajectory of Figure

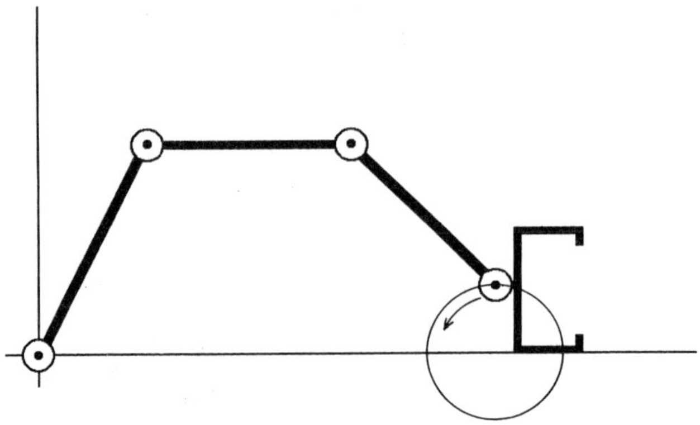

Figure 5: A three bar planar manipulator to trace a circle

6 is of much lower cost than that of Figure 7. In order to better understand this behavior, Figure 8 shows the relationship of these two trajectories when viewed as trajectories in the joint space. We see in this figure a three dimensional rendering of the set of joint angles that satisfy the kinematic constraint $f(\theta) = x(t)$ for some t. Since the manipulator has one degree of redundancy, the set of such joint angles is a set of dimension one greater than the dimension of the set $x(t)$. In this case, it is a surface equivalent to a deformed torus. For fixed t, the set of θ-values corresponding to $x(t)$ is a closed curve that circles the deformed torus about it's larger radius. As time increases from 0 to 1, the manipulator will trace out a curve that circles the torus about the small radius in order to stay on the desired trajectory.

As the figure shows, the lower cost trajectory encircles the torus about the smaller radius at a point where the total distance (in the joint space) that must be traveled is small. The higher cost trajectory encircles the torus about the small radius once, but at the same time, encircles the torus once about the larger radius. Roughly speaking, the manipulator traverses the null space once while performing the motion. It is important to note that the longer trajectory cannot be deformed continuously into the shorter trajectory; we have thus arrived at solutions to the inverse kinematics problem that are in different homotopy classes. Both curves are locally optimal solutions within their homotopy classes.

This example shows that algorithms based on the necessary equations alone will not be able to distinguish solutions in differing homotopy classes, without further refinement. In part to address this problem, and also in part to address the problem of finding computationally efficient ways of solving the boundary value problems associated with global path planning methods, we conclude this section with a report of recent results on applying homotopy continuation methods to the path planning problem.

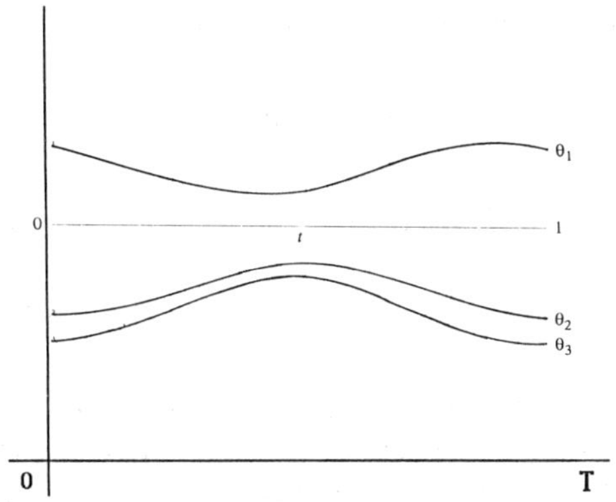

Figure 6: Extremal solution A to the necessary conditions

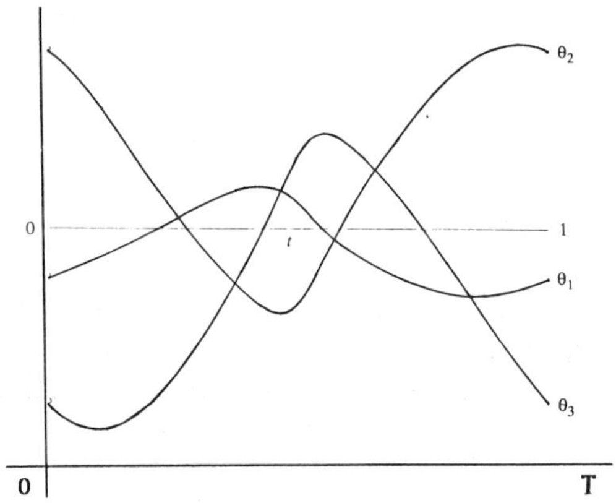

Figure 7: Extremal solution B to the necessary conditions

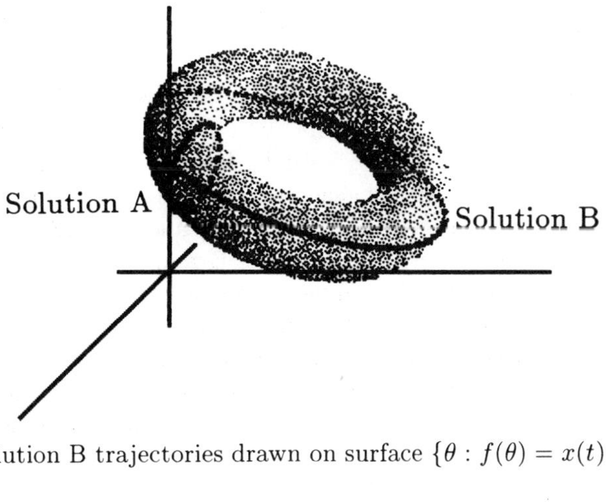

Figure 8: Solution A and solution B trajectories drawn on surface $\{\theta : f(\theta) = x(t), 0 \leq t \leq 1\}$

The specific problem we shall address is that of minimizing the cost functional

$$\int_0^T (\frac{\epsilon}{2}\|\dot{\theta}\|^2 + (1-\epsilon)g(\theta))\,dt \tag{16}$$

subject to the constraint (1) and the periodic boundary conditions

$$\theta(0) = \theta(T), \quad \dot{\theta}(0) = \dot{\theta}(T). \tag{17}$$

We assume the operational space trajectory $x(\cdot)$ is closed (i.e. $x(0) = x(T)$), and the function g appearing in the integrand is a differentiable function of the joint space configuration. (This might be the manipulatability index discussed in the last section, for example.) We also assume for simplicity that the degree of redundancy $n - m = 1$. Equation (13) now takes the form

$$\epsilon\hat{n}\cdot\ddot{\theta} = (1-\epsilon)G(\theta) \tag{18}$$

where

$$G(\theta) \equiv \frac{\partial g}{\partial\theta}(\theta)\cdot\hat{n}(\theta), \tag{19}$$

and $\hat{n}(\theta)$ is the normalized null space vector of the Jacobian $J(\theta)$.

The homotopy parameter ϵ varies between 0 and 1, and when $\epsilon = 0$, the problem reduces to that of extremizing the integral of $g(\theta)$ subject to the constraint. It is not difficult to show that this is equivalent to pointwise optimization of g subject to the constraint (1) at each point θ along the path. Since at $\epsilon = 0$ the motion planning problem is much easier to solve (using, say, the extended Jacobian technique) than the general pathwise optimal problem, we would like to understand whether solutions to the $\epsilon = 0$ problem can in fact be used to initiate a homotopy continuation. We note that this is a potentially troublesome question since the character of the optimization problem (16) changes dramatically as $\epsilon \to 0$. In the limit, the necessary conditions change from the second order differential equation (18) to the algebraic equation $G(\theta) = 0$, so that the homotopy involves a *singular perturbation*.

To bring the issues here into sharper focus we look once again at the manipulator of Example 1.2 together with the local criterion function $g(\theta) = \sin^2\theta_2 + \sin^2\theta_3$. In Example 3.1 we saw that in the case $\ell_j = 1$, $j = 1,2,3$, this gives rise to the constraint $G(\theta) = \theta_2 - \theta_3$. The extended Jacobian technique will generate a joint space trajectory whose point set is a subset of the intersection of the surfaces $\{\theta : G(\theta) = 0\}$ and $\{\theta : f(\theta) = x(t) \text{ for } 0 \le t \le T\}$. For this example and the operational space curve of Example 4.1, the zero locus of G intersects the lifted (inverse) operational space curve locus (the torus) depicted in Figure 8 in curves which encircle the torus about two small equators. Note that of the two extremal trajectories which we have found for the global minimization problem in Example 4.1, only the trajectory we have called Solution A is homotopic to curves associated with the local optimization problem. Solution B, being in a different homotopy class, obviously cannot be continuously deformed into the solution prescribed by the extended Jacobian technique with this criterion function.

At the present writing, it is an active area of research to determine conditions under which solutions $\theta_\epsilon(t)$ of (18)-(17) will converge in a reasonable norm to $\theta_0(t)$, the solution to the algebraic equation $G(\theta) = 0$. While we cannot provide the last word, certain regularity assumptions seem to be essential. We shall thus assume that there are no kinematic singularities in a neighborhood of $\theta_0(t)$. We shall also assume, for simplicity, that the degree of redundancy is one (i.e. $n = m + 1$). Then in a neighborhood of the curve $\theta_0(t)$ there is a continuously defined unit vector $\hat{n}(\theta)$ in the null space of J, and we can prove the following result.

Theorem 4.2 (Martin, [17]) *Given an infinitely differentiable local criterion function $g(\theta)$, suppose that $\theta_0(t)$ is a strict local minimum of g subject to the kinematics constraint $f(\theta(t)) = x(t)$ for a given closed infinitely differentiable operational space curve $x(t)$. Suppose also that there are no kinematic singularities in a neighborhood of $\theta_0(t)$. Then there is an $\epsilon_1 > 0$ and a family of solutions $\theta_\epsilon(t)$ to (18)-(17) such that for $0 < \epsilon < \epsilon_1$, $\lim_{\epsilon\to 0} \|\theta_\epsilon(t) - \theta_0(t)\|_\infty = 0$.*

The proof of this theorem will be carried out through a sequence of lemmas. It will be convenient to rewrite (18) above together with the kinematics constraint (1) as a one dimensional, non-autonomous problem by expressing equation (18) in terms of local coordinates on the manifold $f^{-1}(x(\cdot))$. (The precise conditions under which the inverse image under f of the operational space trajectory is, in fact, a manifold are described in [17].) Note that because we assume $\theta_0(t)$ is the lifting of a smooth operational space path $x(t)$ having a nonzero tangent at each point, it follows that $\theta_0(t)$ has transverse intersection at each point with the null space vector field $\hat{n}(\theta_0(t))$. Figure 9 shows such a θ_0 curve on the example manifold of Figure 8. In the vicinity of $\theta_0(t)$, let $\phi_s(\zeta)$ denote the flow of the vector field $\hat{n}$ with ζ as the initial condition. In other words, $\phi_s(\zeta)$ is the solution to

$$\begin{aligned} \phi_0(\zeta) &= \zeta \\ \frac{d\phi_s}{ds}(\zeta) &= \hat{n}(\phi_s(\zeta)) \end{aligned}$$

Figure 9 depicts such a flow, starting from a ζ on the θ_0 curve. This flow is in fact the nullspace trajectory corresponding to a particular point in the operational space. Since $\hat{n}$ is normalized with respect to some norm on the joint space, the variable s is the distance traveled along the null space trajectory to arrive at the point $\phi_s(\zeta)$. Define $\theta(s,t) \equiv \phi_s(\theta_0(t))$. Since θ_0 is transverse to $\hat{n}$, and by hypothesis there are no kinematic singularities in the neighborhood of $\theta_0(\cdot)$, the pair (s,t) defines a set of of local coordinates on the constraint manifold, as depicted in Figure 10.

By rewriting equation (18) in the (s,t) coordinates, we effectively eliminate the constraint from (18). First compute $\hat{n}(\theta) \cdot \ddot{\theta}$ in the (s,t) coordinates. Under the hypothesis of the theorem, there is enough smoothness in the quantities in question to write

$$\frac{d\theta}{dt}(s,t) = \hat{n}(\theta(s,t)) \cdot \dot{s} + \frac{\partial\theta}{\partial t}(s,t)$$

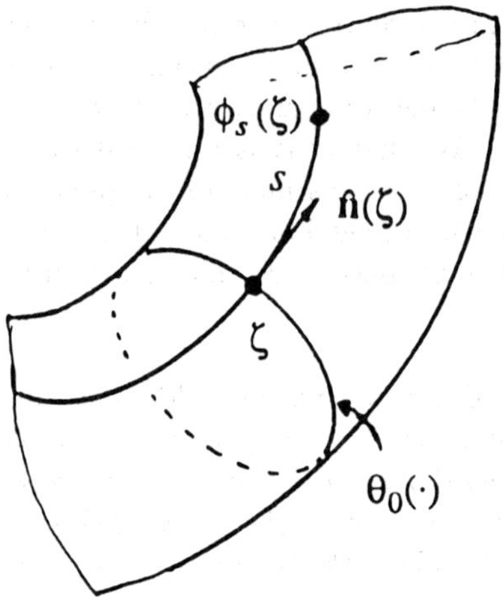

Figure 9: The curve $\theta_0(\cdot)$ and the flow of $\hat{n}$.

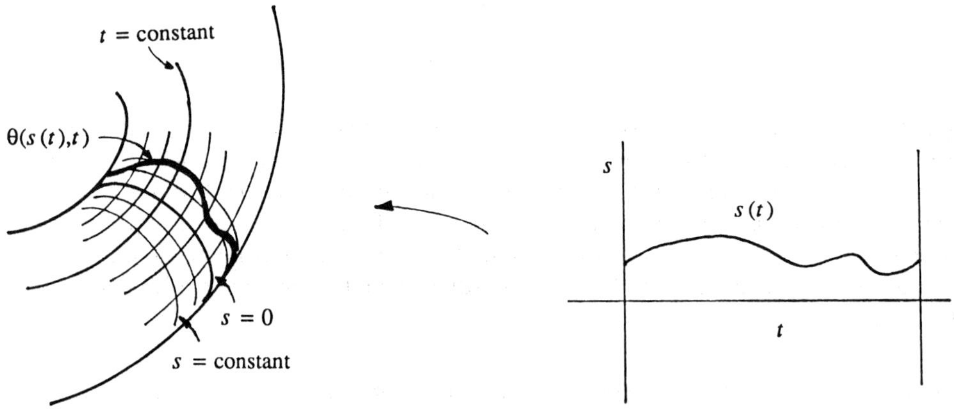

Figure 10: The (s,t) coordinates defined by $\theta_0(\cdot)$ and the flow of $\hat{n}$ on the constraint manifold

and

$$\frac{d^2\theta}{dt^2}(s,t) = 2\dot{\hat{n}} \cdot \dot{s} + \hat{n} \cdot \ddot{s} + \frac{\partial^2\theta}{\partial t^2}(s,t),$$

where $\dot{\hat{n}} = \frac{d}{dt}\hat{n}(s(t),t)$. Thus

$$\epsilon\hat{n} \cdot \ddot{\theta} = \epsilon\ddot{s} + \epsilon\hat{n} \cdot \frac{\partial^2\theta}{\partial t^2}$$

where we have made use of the fact that $\hat{n}$ is a unit vector so that $\hat{n} \cdot \hat{n} = 1$ and $\hat{n} \cdot \dot{\hat{n}} = 0$. Hence we may rewrite (18) as

$$\epsilon\ddot{s} = -\epsilon f_1(s(t),t) + (1-\epsilon)f_2(s(t),t)$$

where

$$f_1(s,t) = -\epsilon\hat{n}(\theta(s,t)) \cdot \frac{\partial^2\theta}{\partial t^2}(s,t) \quad \text{and} \quad f_2(s,t) = G(\theta(s,t)).$$

Since we are interested in the case where ϵ approaches zero, it will be convenient to replace ϵ by $\epsilon^2/(1+\epsilon^2)$ which allows us to rewrite the above equation as

$$\epsilon^2\ddot{s} \;=\; -\epsilon^2 f_1(s,t) \;+\; f_2(s,t)\;. \tag{20}$$

If we choose $\theta_0(t)$ to be a curve that minimizes $g(\theta(t))$ at each t subject to the kinematics constraint, then Theorem 2.3 implies that $G(\theta_0(t)) \equiv 0$. A straightforward argument further shows that $\frac{\partial G}{\partial\theta}(\theta_0(t)) \cdot \hat{n}(\theta_0(t)) > 0$. This implies that at $s = 0$ the term f_2 is such that

$$\begin{aligned} f_2(0,t) \;&=\; 0 \\ \frac{\partial f_2}{\partial s}(0,t) \;&>\; 0 \end{aligned}$$

In this recast form, the problem under consideration is to determine circumstances under which solutions to (20) exist, and converge in an appropriate sense to θ_0 as $\epsilon \to 0$.

Remark 4.1 Our discussion will differ from what is found in standard sources (e.g. [24]), in that the first-orderized version of the second order problem above fails to meet the technical conditions usually imposed on the leading term in the asymptotic expansion of the Jacobian. These conditions exclude turning points, requiring that the partial of the right hand side of the equation with respect to $\dot{s}$ be bounded away from zero. In the case we are considering, this partial is zero over the entire domain of t.

Remark 4.2 In the inequalities to be developed in the following lemmas, all norms on functions are assumed to be *sup* norms unless otherwise indicated.

Lemma 4.1 *Suppose $a(t)$, $b(t)$ are twice differentiable and periodic of period $T > 0$ on $[0, T]$, with $a(t) > 0$. Then there is an $\epsilon_1 > 0$ such that for each $0 < \epsilon < \epsilon_1$ the differential equation*

$$\epsilon^2 \ddot{x} = a(t)x + \epsilon^2 b(t)$$

has a solution $x(t)$ which is periodic of period T and is such that $\|x\| < \epsilon^2 D\|b\|$, and $\|\dot{x}\| < \epsilon B\|b\|$ for some $B, D > 0$. Here, $\|\cdot\|$ is the sup norm.

Proof: We "first-orderize" the system

$$\epsilon^2 \ddot{x} = a(t)x + \epsilon^2 b(t)$$

to obtain, for $\epsilon > 0$ the vector system of equations

$$\dot{\mathbf{x}} = A(t)\mathbf{x} + \mathbf{b}(t)$$

where

$$\mathbf{x} = \begin{bmatrix} x \\ \dot{x} \end{bmatrix}$$

$$\mathbf{b} = \begin{bmatrix} 0 \\ b(t) \end{bmatrix}$$

$$A(t) = \begin{bmatrix} 0 & 1 \\ a(t)/\epsilon^2 & 0 \end{bmatrix}.$$

We diagonalize $A(t)$ at each instant of time via the change of variables

$$\mathbf{y}(t) = U^{-1}(t)\mathbf{x}(t)$$

$$U(t) = \begin{bmatrix} 1 & 1 \\ \sqrt{a}/\epsilon & -\sqrt{a}/\epsilon \end{bmatrix}$$

$$U^{-1}(t) = \begin{bmatrix} 1/2 & \epsilon/(2\sqrt{a}) \\ 1/2 & -\epsilon/(2\sqrt{a}) \end{bmatrix}$$

The differential equation then takes the form

$$\dot{\mathbf{y}} = \Lambda(t)\mathbf{y} + \dot{U}^{-1}(t)U(t)\mathbf{y} + U^{-1}(t)\mathbf{b}$$

where

$$\Lambda(t) = \begin{bmatrix} \sqrt{a}/\epsilon & 0 \\ 0 & -\sqrt{a}/\epsilon \end{bmatrix}.$$

Next, use is made of the variation of parameters formula appropriate to the periodic boundary conditions of this problem:

$$\mathbf{y}(t) = (I - \Phi(t, t-T))^{-1} \int_{t-T}^{t} \Phi(t,\tau)\, \dot{U}^{-1}(\tau)U(\tau)\mathbf{y}(\tau) + U^{-1}(\tau)\mathbf{b}(\tau)\, d\tau$$

where

$$\Phi(t,\tau) = \begin{bmatrix} e^{\frac{1}{\epsilon}\alpha(t,\tau)} & 0 \\ 0 & e^{-\frac{1}{\epsilon}\alpha(t,\tau)} \end{bmatrix}$$

$$\alpha(t,\tau) = \int_\tau^t \sqrt{a(s)}ds.$$

We expect from Floquet theory and the fact that Φ is diagonal that $\Phi(t, t-T)$ is independent of t. If we define

$$\bar{\alpha} = \int_0^T \sqrt{a(s)}ds$$

we see that

$$(I - \Phi(t,t-T))^{-1} = \begin{bmatrix} (1-e^{\bar{\alpha}/\epsilon})^{-1} & 0 \\ 0 & (1-e^{-\bar{\alpha}/\epsilon})^{-1} \end{bmatrix}.$$

Noting that $\alpha(t,\tau) = \bar{\alpha} - \alpha(\tau, t-T)$, we see that

$$(I - \Phi(t,t-T))^{-1}\Phi(t,\tau) = (1-e^{-\bar{\alpha}/\epsilon})^{-1} \begin{bmatrix} -e^{-\alpha(\tau,t-T)/\epsilon} & 0 \\ 0 & e^{-(\bar{\alpha}-\alpha(\tau,t-T))/\epsilon} \end{bmatrix}.$$

If we define $\Psi(\tau, t-T)$ to be the quantity above, the variation of parameters formula becomes

$$\mathbf{y}(t) = \int_{t-T}^t \Psi(\tau,t-T)\, \dot{U}^{-1}(\tau)U(\tau)\mathbf{y}(\tau) + U^{-1}(\tau)\mathbf{b}(\tau)\, d\tau.$$

We note that $\Psi(\tau, t-T)$ remains bounded on $t - T < \tau < t$ as we vary $\epsilon > 0$, in contrast with $\Phi(t,\tau)$.

In order to obtain bounds on $\mathbf{x}$, we change variables back to $\mathbf{x} = U\mathbf{y}$:

$$\mathbf{x}(t) = U(t)\int_{t-T}^t \Psi(\tau,t-T)\, \dot{U}^{-1}(\tau)\mathbf{x}(\tau) + U^{-1}(\tau)\mathbf{b}(\tau)\, d\tau.$$

Next we examine the two terms in this integral. If we define

$$c(t) = \frac{\dot{a}(t)}{2a(t)}$$

then we have

$$\dot{U}^{-1}(\tau) = c(\tau)\begin{bmatrix} 0 & -\epsilon/(2\sqrt{a(\tau)}) \\ 0 & \epsilon/(2\sqrt{a(\tau)}) \end{bmatrix}.$$

We compute

$$U(t)\Psi(\tau,t-T)\dot{U}^{-1}(\tau) = c(\tau)\begin{bmatrix} 0 & \frac{\epsilon}{\sqrt{a(\tau)}}K_1(\tau,t) \\ 0 & \frac{\sqrt{a(t)}}{\sqrt{a(\tau)}}K_2(\tau,t) \end{bmatrix},$$

and

$$U(t)\Psi(\tau,t-T)U^{-1}(\tau) = \begin{bmatrix} -K_2(\tau,t) & -\frac{\epsilon}{\sqrt{a(\tau)}}K_1(\tau,t) \\ -\sqrt{\frac{a(t)}{\epsilon}}K_1(\tau,t) & -\frac{\sqrt{a(t)}}{\sqrt{a(\tau)}}K_2(\tau,t) \end{bmatrix}$$

where

$$K_1(\tau,t) = \frac{k}{2}(e^{-\alpha(\tau,t-T)/\epsilon} + e^{-(\bar{\alpha}-\alpha(\tau,t-T)/\epsilon})$$

$$K_2(\tau, t) = \frac{k}{2}(e^{-\alpha(\tau,t-T)/\epsilon} - e^{-(\bar{\alpha}-\alpha(\tau,t-T)/\epsilon})$$
$$k = (1 - e^{-\bar{\alpha}/\epsilon})^{-1}.$$

With this, we can write the integral equation as

$$x(t) = \epsilon \int_{t-T}^{t} \frac{K_1(\tau, t)c(\tau)}{\sqrt{a(\tau)}} \dot{x}(\tau)d\tau - \epsilon \int_{t-T}^{t} \frac{K_1(\tau, t)}{\sqrt{a(\tau)}} b(\tau)d\tau$$

and

$$\dot{x}(t) = \sqrt{a(t)} \int_{t-T}^{t} \frac{K_2(\tau, t)c(\tau)}{\sqrt{a(\tau)}} \dot{x}(\tau)d\tau - \sqrt{a(t)} \int_{t-T}^{t} \frac{K_2(\tau, t)}{\sqrt{a(\tau)}} b(\tau)d\tau ,$$

or, collecting terms, as

$$x(t) = \epsilon \int_{t-T}^{t} \frac{K_1(\tau, t)}{\sqrt{a(\tau)}} (c(\tau)\dot{x}(\tau) - b(\tau))\, d\tau$$

and

$$\dot{x}(t) = \sqrt{a(t)} \int_{t-T}^{t} \frac{K_2(\tau, t)}{\sqrt{a(\tau)}} (c(\tau)\dot{x}(\tau) - b(\tau))\, d\tau .$$

We will write these two integral equations in terms of the operators F_1 and F_2:

$$F_1(z(\cdot)) \equiv \epsilon \int_{t-T}^{t} \frac{K_1(\tau, t)}{\sqrt{a(\tau)}} z(\tau)d\tau$$

$$F_2(z(\cdot)) \equiv \sqrt{a(t)} \int_{t-T}^{t} \frac{K_2(\tau, t)}{\sqrt{a(\tau)}} z(\tau)d\tau.$$

The integral equations can then be written

$$x = F_1(c \cdot \dot{x} - b)$$
$$\dot{x} = F_2(c \cdot \dot{x} - b).$$

To show that a solution to these equations exists, we proceed as usual to show that the solution is the limit of a sequence of functions obtained from Picard iteration. We note that a solution to the second equation is sufficient, as the solution to the first may be determined from it by a single integration. Define

$$\dot{x}_0 = 0$$
$$\dot{x}_{n+1} = F_2(c \cdot \dot{x}_n - b).$$

We see that

$$\dot{x}_{n+1} - \dot{x}_n = F_2(c \cdot (\dot{x}_n - \dot{x}_{n-1})).$$

Next, we wish to obtain bounds on the integrals above.

Since we have assumed $a(t) > 0$, the function $\alpha(\tau, t-T)$ appearing in $K_i(\tau,t)$ is monotonically increasing, and therefore has an inverse function, which we denote as $\tau(\alpha)$:

$$\alpha = \int_{t-T}^{\tau(\alpha)} \sqrt{a(s)}ds.$$

We can thus change the variable of integration to α with $d\alpha = \sqrt{a(\tau)}d\tau$:

$$\epsilon \int_{t-T}^{t} \frac{K_1(\tau,t)}{\sqrt{a(\tau)}} z(\tau)d\tau = \epsilon \int_0^{\bar{\alpha}} \frac{K_1(\tau(\alpha),t)}{a(\tau(\alpha))} z(\tau(\alpha))d\alpha$$

$$\sqrt{a(t)} \int_{t-T}^{t} \frac{K_2(\tau,t)}{\sqrt{a(\tau)}} z(\tau)d\tau = \sqrt{a(t)} \int_0^{\bar{\alpha}} \frac{K_2(\tau(\alpha),t)}{a(\tau(\alpha))} z(\tau(\alpha))d\alpha.$$

In the sup norm,

$$\|z(t)\| = \sup_{0<t<T} |z(t)| = \sup_{0<\alpha<\bar{\alpha}} |z(\tau(\alpha))|,$$

we can bound the integral equations as

$$\|F_1(z(\cdot))\| < \epsilon\|z/a\| \int_0^{\bar{\alpha}} |\tilde{K}_1(\alpha)|d\alpha$$

$$\|F_2(z(\cdot))\| < \|\sqrt{a}\| \; \|z/a\| \int_0^{\bar{\alpha}} |\tilde{K}_2(\alpha)|d\alpha$$

where

$$\tilde{K}_1(\alpha) = \frac{k}{2}(e^{-\alpha/\epsilon} + e^{-(\bar{\alpha}-\alpha)/\epsilon})$$

$$\tilde{K}_2(\alpha) = \frac{k}{2}(e^{-\alpha/\epsilon} - e^{-(\bar{\alpha}-\alpha)/\epsilon}).$$

Since $|\tilde{K}_2| < |\tilde{K}_1|$ and $|\tilde{K}_1| = \tilde{K}_1$ at each α, we can bound the integrals above by

$$\epsilon = \int_0^{\bar{\alpha}} \tilde{K}_1(\alpha)d\alpha.$$

The bounds on the integral operators are then

$$\|F_1(z)\| < \epsilon^2\|z/a\|$$

$$\|F_2(z)\| < \epsilon\|\sqrt{a}\|\|z/a\|.$$

We see then, that

$$\|\dot{x}_{n+1} - \dot{x}_n\| < \|F_2(c\cdot(\dot{x}_n - \dot{x}_{n-1}))\| < \epsilon\|\sqrt{a}\|\|c/a\|\|\dot{x}_n - \dot{x}_{n-1}\|$$

or

$$\|\dot{x}_{n+1} - \dot{x}_n\| < \epsilon M_{21}\|\dot{x}_n - \dot{x}_{n-1}\|$$

where M_{21} is a function only of a and its derivatives, and is bounded by our hypotheses on a. We see that the sequence $\dot{x}_n$ is a Cauchy sequence for $\epsilon < \epsilon_0$, $\epsilon_0 = M_{21}^{-1}$. The sequence then converges to a limit point as the set of continuous periodic functions $C_p^0[0,T]$ is complete in the sup norm. The sequence x_n thus converges.

This implies that $\dot{x}$ and hence x satisfying the differential equation for $0 < \epsilon < \epsilon_0$ exists. Next, we wish to find bounds on these solutions. Define

$$\begin{aligned} \dot{x}_0 &= 0 \\ \dot{x}_{n+1} &= F_2(c \cdot \dot{x}_n - b). \end{aligned}$$

Applying the bounds above to the sequence, we see that

$$\begin{aligned} \|\dot{x}_1\| &< \epsilon\|\sqrt{a}\|\|b/a\| \\ &< \epsilon M_{22}\|b/a\|. \\ \|\dot{x}_2\| < \epsilon\|\sqrt{a}\|\|(c\cdot\dot{x}_1 - b)/a\| &< \epsilon\|\sqrt{a}\|(\|c/a\|\|\dot{x}_1\| + \|b/a\|) \\ &< \epsilon(M_{21}\|\dot{x}_1\| \ + \ M_{22}\|b/a\|) \\ &< (\epsilon^2 M_{21} + \epsilon)M_{22}\|b/a\|. \end{aligned}$$

Continuing in this fashion, we see that

$$\|\dot{x}_n\| \ < \ \epsilon\sum_{i=0}^{n-1}(\epsilon M_{21})^i M_{22}\|b/a\|.$$

This bound on $\dot{x}_n$ converges to $\epsilon(1-\epsilon M_{21})^{-1}M_{22}\|b/a\|$. This implies that we can chose B and ϵ_1 so that

$$\|\dot{x}\| < \epsilon B\|b/a\|$$

for all $\epsilon < \epsilon_1$. We have similar bounds on $\|x\|$, as $\|x\| < \epsilon^2(\|c/a\|\|\dot{x}\| \ + \ \|b/a\|)$:

$$\|x\| \ < \ \epsilon^2(\epsilon C \ + \ E)\|b\|.$$

Finally, we take D to be $\epsilon_1 C + E$ to prove the lemma. □

Next, we will slightly extend the lemma above to include an ϵ^2 perturbation of $a(t)$. For simplicity, we will prove a simple extension of Lemma 4.1 using a variation of Picard iteration that will be used to handle the nonlinear case.

Lemma 4.2 *Suppose $a_0(t)$, $a_1(t)$, and $b(t)$ are twice differentiable and periodic of period $T > 0$. If $a_0(t) > 0$, the differential equation*

$$\epsilon^2\ddot{x} \ = \ a_\epsilon(t)x \ + \ \epsilon^2 b(t),$$

where $a_\epsilon(t) \ = \ a_0(t) \ + \ \epsilon^2 a_1(t)$, has a periodic solution $x(t)$ for $0 < \epsilon < \epsilon_1$ such that $\|x\| \ < \ \epsilon^2 D\|b\|$ and $\|\dot{x}\| \ < \ \epsilon B\|b\|$ for real numbers $B, D > 0$.

Proof: Consider the following sequence of differential equations:

$$\epsilon^2 \ddot{x}_n = a_0(t)x_n + \epsilon^2(a_1(t)x_{n-1} + b(t))$$

where $x_0 \equiv 0$. First, we consider the differences, $v_n = x_n - x_{n-1}$. These satisfy

$$v_1 = x_1$$

and

$$\epsilon^2 \ddot{v}_n = a_0(t)v_n + \epsilon^2 a_1(t)v_{n-1}.$$

This equation has a solution v_n such that for B and D as found in Lemma 4.1

$$\|v_n\| < \epsilon^2 D\|a_1\|\|v_{n-1}\|$$

$$\|\dot{v}_n\| = \epsilon B\|a_1\|\|v_{n-1}\|.$$

We see that for ϵ small enough, $\|v_n\| \to 0$ as $n \to \infty$, and $\|\dot{v}_n\| \to 0$ as $n \to \infty$. As in Lemma 4.1, this means that the x_n converge to a differentiable function. Next, we find bounds on the x_n: If we consider

$$\epsilon^2 \ddot{x}_n = a_0(t)x_n + \epsilon^2(a_1(t)x_{n-1} + b(t))$$

we may use the bounds of Lemma 4.1 to obtain

$$\|x_n\| < \epsilon^2 D\|a_1\|\|x_{n-1}\| + \epsilon_2 D\|b\|$$

and

$$\|\dot{x}_n\| < \epsilon_2 B(\|a_1\|\|x_{n-1}\| + \|b\|)$$

As in the proof of Lemma 4.1, the first sequence of inequalities is bounded above by a geometric series, so that

$$\lim_{n\to\infty} \|x_n\| < \frac{\epsilon^2 D\|b\|}{(1 - \epsilon^2 D\|a_0\|)}$$

For small enough ϵ, this provides the bound

$$\|x\| < \epsilon_2 \tilde{D}\|b\|$$

Using this fact in the second series, we have

$$\lim_{n\to\infty} \|\dot{x}_n\| < \epsilon B(\epsilon_2 \tilde{D}\|a_1\| + 1)\|b\|$$

Again, for small enough epsilon, we may write $\|\dot{x}\| < \epsilon \tilde{B}\|b\|$. □

Lemma 4.3 *Given $a_\epsilon(t)$, as in Lemma 4.2, and $b(t)$ twice differentiable and periodic of period T, the differential equation*

$$\epsilon^2 \ddot{x} = a_\epsilon(t)x + b(t)$$

has a periodic solution $x(t)$ for each $0 < \epsilon < \epsilon_1$ such that $\|x\|_* < K\|b\|_*$, where $K > 0$ and $\|\cdot\|_*$ is the norm

$$\|f\|_* = \|\ddot{f}\| + \|\dot{f}\| + \|f\|$$

Proof: First, we put the differential equation into a form suitable for the application of Lemma 4.2. Define $x_0 \equiv -b(t)/a_0(t)$ and $\delta x \equiv x - x_0$. This puts the equation into the form

$$\epsilon_2 \ddot{\delta x} = a_\epsilon(t)\delta x + \epsilon^2(a_1(t)x_0 - \ddot{x}_0).$$

We compute

$$\ddot{x}_0 = -\frac{1}{a_0}\left[\ddot{b} - 2\dot{b}\frac{\dot{a}_0}{a} + 2b\left(\frac{\dot{a}_0}{a_0}\right)^2 + b\ddot{a}_0\right].$$

From this, we see that

$$\| - \ddot{x}_0 + a_1 x_0\| < K_1\|b\|_*$$

for some K_1 depending only on $a_0(t)$ and its derivatives. From the differential equation, we can bound $\delta\ddot{x}$

$$\|\delta\ddot{x}\| < \frac{1}{\epsilon^2}\|a\|\|\delta x\| + \|a_1 x_0 - \ddot{x}_0\|$$

where $\delta x = x - x_0$ plays the role of the unknown variable as in the proof of Lemma 4.1.

From the bounds provided by lemma 1, we have

$$\|\delta x\|_* = \|\delta\ddot{x}\| + \|\delta\dot{x}\| + \|\delta x\|$$

$$< (\|a\|D\|a_1 x_0 - \ddot{x}_0\| + \|a_1 x_0 - \ddot{x}_0\|) + \epsilon B\|a_1 x_0 - \ddot{x}_0\| + \epsilon^2 D\|a_1 x_0 - \ddot{x}_0\|$$

$$< K_2\|a_1 x_0 - \ddot{x}_0\| < K_3\|b\|_*.$$

By the triangle inequality,

$$\|x\|_* < \|x_0\|_* + \|\delta x\|_*.$$

Since $\|x_0\|_* = \|b/a_0\|_*$ we have $\|x\|_* < \|b/a_0\|_* + K_3\|b\|_*$, or finally, $\|x\|_* < K\|b\|_*$. □

Lemma 4.4 *Suppose $g(x,t)$ is such that for all $0 \le t \le T$*

$$g(0,t) = 0\ , \quad \frac{dg}{dx}(0,t) = 0$$

and that furthermore for some $\eta > 0$ and all $|x| < \eta$ the following are bounded (uniformly in $t \in [0,T]$):

$$\|\frac{d^2g}{dx^2}(x,t)\| < \infty\ , \quad \|\frac{d^3g}{dx^3}(x,t)\| < \infty\ ,$$

$$\|\frac{d^2g}{dtdx}(x,t)\| < \infty\ , \quad \|\frac{d^3g}{dt^2dx}(x,t)\| < \infty\ ,$$

$$\|\frac{dg}{dt}(x,t)\| < \infty\ , \quad \|\frac{d^2g}{dt^2}(x,t)\| < \infty.$$

Then there exists a $\delta_1 > 0$ and a continuous positive function $M(\delta)$ defined for $0 < \delta < \delta_1$ such that if $\|x_i\| < \delta$ for $i = 1,2$, then

$$\|g(x_1) - g(x_2)\|_* < M(\delta)\|x_1 - x_2\|_*,$$

and $M(\delta) > 0$ satisfies $\lim_{\delta\to 0} M(\delta) = 0$.

Proof: We write the difference $g(x_1, t) - g(x_2, t)$ as

$$g(x_1, t) \ - \ g(x_2, t) \ = \ \int_{x_2}^{x_1} \frac{dg}{dx} dx.$$

Defining $h = \frac{dg}{dx}$ and changing the variable of integration to s where $x = sx_1 + (1-s)x_2$ this becomes

$$g(x_1, t) \ - \ g(x_2, t) \ = \ \int_0^1 (x_1 \ - \ x_2) h(sx_1 \ + \ (1-s)x_2, t) ds \ .$$

If we set $M_0 = \sup_{\substack{0 \le s \le 1 \\ t}} h(sx_1 + (1-s)x_2, t)$, then

$$|g(x_1, t) - g(x_2, t)| \ \le \ M_0 |x_1 - x_2|.$$

Next, we examine the first derivative:

$$\frac{d}{dt}[\ g(x_1, t) \ - \ g(x_2, t)] \ = \ \int_0^1 [(\dot{x}_1 \ - \ \dot{x}_2) h(sx_1 \ + \ (1-s)x_2, t)] \ ds$$

$$+ \int_0^1 [(x_1 \ - \ x_2) h_x(sx_1 \ + \ (1-s)x_2)(s\dot{x}_1 \ + \ (1-s)\dot{x}_2)] \ ds$$

$$+ \int_0^1 [(x_1 \ - \ x_2) h_t(sx_1 \ + \ (1-s)x_2)] ds.$$

If we define

$$M_1 \ = \ \sup_{\substack{0 \le s \le 1 \\ t}} \ [h_x(sx_1 + (1-s)x_2, t)(s\dot{x}_1 + (1-s)\dot{x}_2) \ + \ h_t(sx_1 + (1-s)x_2, t)]$$

then we see that

$$|\dot{g}(x_1, t) - \dot{g}(x_2, t)| \ < \ M_0 |\dot{x}_1 - \dot{x}_2| \ + \ M_1 |x_1 - x_2|$$

Next, we compute the second derivative:

$$\frac{d^2}{dt^2} [g(x_1, t) \ - \ g(x_2, t)] \ = \ \int_0^1 (\ddot{x}_1 \ - \ \ddot{x}_2) h \ ds$$

$$+ \int_0^1 (\dot{x}_1 \ - \ \dot{x}_2) \ 2[(s\dot{x}_1 \ + \ (1-s)\dot{x}_2) h_x + h_t] \ ds$$

$$+ \int_0^1 (x_1 - x_2) \ [(s\dot{x}_1 + (1-s)\dot{x}_2)^2 h_{xx} + 2(s\dot{x}_1 + (1-s)\dot{x}_2) h_{xt} + h_{tt} \ + \ (s\ddot{x}_1 + (1-s)\ddot{x}_2) h_x] \ ds$$

where we have supressed the argument of h and its derivatives. Defining

$$M_2 \ = \ \sup_{\substack{0 \le s \le 1 \\ t}} \ [(s\dot{x}_1 + (1-s)\dot{x}_2)^2 h_{xx} + 2(s\dot{x}_1 + (1-s)\dot{x}_2) h_{xt} + h_{tt} \ + \ (s\ddot{x}_1 + (1-s)\ddot{x}_2) h_x]$$

we see that

$$|\ddot{g}(x_1) - \ddot{g}(x_2)| \ < \ M_0|\ddot{x}_1 - \ddot{x}_2| \ + \ 2M_1|\dot{x}_1 - \dot{x}_2| \ + \ M_2|x_1 - x_2|$$

This gives us

$$\begin{aligned} \|g(x_1) - g(x_2)\|_* \ &< \ (M_0 + M_1 + M_2)|x_1 - x_2| \\ &+ (M_0 + 2M_1)|\dot{x}_1 - \dot{x}_2| \ + \ M_0|\ddot{x}_1 - \ddot{x}_2| \ . \end{aligned}$$

If $\|x_1\|_*$ and $\|x_2\|_*$ are less that δ, we can bound M_0, M_1, and M_2. This gives us

$$\|g(x_1) - g(x_2)\|_* \ < \ M\|x_1 - x_2\|_*$$

for some finite M. Further, M goes to zero as $\|x_1\|_*$ and $\|x_2\|_*$, since M_0 , M_1 and M_2 all go to zero with $\|x_1\|_*$ and $\|x_2\|_*$. □

Theorem 4.3 *Let $f_1(x,t)$ and $f_2(x,t)$ be four times differentiable, periodic in t, with*

$$f_2(0,t) \ = \ 0$$

Further, assume that

$$\|f_1(0,t)\|_* \ < \ \infty$$

and

$$\frac{\partial f_2}{\partial x}(0,t) \ > \ 0$$

Under these assumptions, the equation

$$\epsilon^2 \ddot{x} \ = \ -\epsilon^2 f_1(x,t) \ + \ f_2(x,t)$$

has periodic solutions x_ϵ for $0 < \epsilon < \epsilon_1$, where $x_\epsilon \to 0$ uniformly as $\epsilon \to 0$.

Proof: We expand the differential equation about $x \equiv 0$. The differential equation becomes

$$\epsilon^2 \ddot{x} \ = \ -\epsilon^2 f_1(0,t) \ + \ \left(\frac{\partial f_2}{\partial x}(0,t) - \epsilon^2 \frac{\partial f_1}{\partial x}(0,t)\right) x \ + \ g_2(x,t) - \epsilon^2 g_1(x,t).$$

The rightmost terms are the remainders of the first order expansions of f_2 and $-\epsilon^2 f_1$ about $x \equiv 0$, respectively. If we set $g_\epsilon(x,t) \ = \ g_2(x,t) - \epsilon^2 g_1(x,t)$, $a_\epsilon \ = \ \frac{\partial f_2}{\partial x}(0,t) - \epsilon^2 \frac{\partial f_1}{\partial x}(0,t)$, and $b(t) \ = \ f_1(0,t)$, we have the differential equation in the form

$$\epsilon^2 \ddot{x} \ = \ -\epsilon^2 b(t) \ + \ a_\epsilon(t)x \ + \ g_\epsilon(x,t).$$

We now make use of Lemmas 4.2 and 4.3 and a variant of Picard iteration to show the existence of a solution.

First, we restrict the range of ϵ so that $a_\epsilon > 0$. Choose ϵ_1 so that $0 < \epsilon < \epsilon_1$ implies $a_\epsilon(t) > 0$ for all t.

Define the sequence of functions u_n by the recursion

$$\epsilon^2 \ddot{u}_n = -\epsilon^2 b(t) + a_\epsilon(t) u_n + g_\epsilon(u_{n-1}, t)$$

where $u_0 \equiv 0$. We will later require that all $\|u_n\|_*$ be less that some $\delta > 0$. We will establish this by induction. Suppose that the solution u_{n-1} exists. We can then use Lemma 4.3, with $-\epsilon^2 b(t) + g_\epsilon(u_{n-1}, t)$ playing the role of $b(t)$ in that lemma, to show that

$$\|u_n\|_* < K\epsilon^2 \|b(t)\|_* + K\|g_\epsilon(u_{n-1}, t)\|_*.$$

If $\|u_{n-1}\|_* < \delta$ we may use Lemma 4.4 to bound the right hand term:

$$\|u_n\|_* < K\epsilon^2 \|b(t)\|_* + KM\|u_{n-1}\|_*.$$

Since $M \to 0$ as $\delta \to 0$, we can find δ small enough that $\|u_{n-1}\|_* < \delta$ implies that $KM < 1/2$. At the same time, we can find $\epsilon_2 < \epsilon_1$ small enough that $K{\epsilon_2}^2\|b(t)\|_* < \frac{1}{4}\delta$. Thus, when $\epsilon < \epsilon_2$ and $\|u_{n-1}\|_* < \delta$ we have

$$\|u_n\|_* < \frac{1}{4}\delta + \frac{1}{2}\|u_{n-1}\|_*$$

This implies that $\|u_n\|_* < \delta$ as well. Since $u_0 = 0$ we have $\|u_1\|_* < \frac{1}{4}\delta$. Therefore, when $\epsilon < \epsilon_2$, $\|u_n\|_* < \delta$ for all n by induction. Thus, by choice of ϵ_2, we can guarantee that all of the u_n's are small. In fact, the sequence above is a geometric sequence bounded by

$$\|u_n\|_* < 2K{\epsilon_2}^2\|b(t)\|_* .$$

Now, though we have shown that the u_n's are bounded, we must show that they converge. Consider the differences $v_n \equiv u_n - u_{n-1}$. The v_n must satisfy

$$\epsilon^2 \ddot{v}_n = a_\epsilon(t) v_n + g_\epsilon(u_{n-1}, t) - g_\epsilon(u_{n-2}, t)$$

By Lemma 4.3 the equation above has solutions v_n where

$$\|v_n\|_* < K\|g_\epsilon(u_{n-1}, t) - g_\epsilon(u_{n-2}, t)\|_*$$

We can again apply Lemma 4.4 to bound the rightmost term:

$$\|g_\epsilon(u_n, t) - g_\epsilon(u_{n-1}, t)\|_* < M\|u_{n-1} - u_{n-2}\|_*$$

so that

$$\|v_n\|_* < KM\|v_{n-1}\|_*$$

In order to show convergence, we will establish conditions under which $KM < 1$. Recall that in lemma 3, $M \to 0$ as $\|u_n\|_*$ and $\|u_{n-1}\|_*$ do. Thus, there exists some $\delta > 0$ so that $\|u_n\|_* < \delta$

for all n implies that $KM < 1$. Since we have shown that we can choose ϵ_2 so that $\|u_n\|_* < \delta$ for all n, there is some $\epsilon_2 > 0$ such that $MK < 1$.

This establishes that the sequence u_n is Cauchy in the $\|\cdot\|_*$ norm. As the space of twice differentiable functions is complete in this norm, the u_n's converge to a twice differentiable function that is a solution to the differential equation

$$\epsilon^2 \ddot{x} = -\epsilon^2 f_1(x,t) + f_2(x,t) .$$

Since

$$\|u_n\|_* < 2K\epsilon^2 \|b(t)\|_* ,$$

we have

$$\|x\|_* < 2K\epsilon^2 \|f_1(0,t)\|_* ,$$

for $0 < \epsilon < \epsilon_2$. Thus, the solution to the equation of the theorem $x(t)$ approaches 0 uniformly as $\epsilon \to 0$. □

Proof of Theorem 4.2: Note that under the assumptions of Theorem 4.2 f_1 and f_2 in equation (20) satisfy the hypotheses of Theorem 4.3. This observation proves Theorem 4.2. □

Many questions regarding pathwise optimal resolution of kinematic redundancy remain at this writing. While the approach we have described may be thought of as being "global" with respect to a given operational space path, it remains unknown how the results of the method may be expected to change as operational space paths are varied. The local methods of Section 3 were seen to inevitably involve algorithmic singularities, and it seems unlikely that the repeatable path otimal methods can be free of such pathology. Indeed, the existence of multiple, nonhomotopic, locally optimal liftings of operational space paths suggests that as problem parameters are varied, there will be bifurcations of such solutions. There may be much work remaining to understand such phenomena.

References

[1] Baillieul, J., 1985. "Kinematic Programming Alternatives for Redundant Manipulators," *IEEE International Conference on Robotics and Automation*, St. Louis MO, March 25-28, pp. 722-728.

[2] Baillieul, J., 1987. "A Constraint Oriented Approach to Inverse Problems for Kinematically Redundant Manipulators," *IEEE International Conference on Robotics and Automation*, Raliegh, NC, March 31-April 3, pp. 1827-1833.

[3] Baillieul, J., 1986. "Avoiding Obstacles and Resolving Kinematic Redundancy," *Proc. IEEE Int. Conf. on Robotics and Automation*, San Francisco, CA, April 7-10, pp. 1698-1704.

[4] Baillieul, J., Hollerbach, J.M., and Brockett, R.W., 1984: "Programming and Control of Kinematically Redundant Manipulators," *Proc. of 1984 IEEE Conference on Decision and Control*, Las Vegas, Nevada, pp. 768-774.

[5] Baker, D. R. and Wampler, C. W. 1988. "On the Inverse Kinematics of Redundant Manipulators," *International Journal of Robotics Research*, vol. 7, no. 2, pp. 3-21.

[6] Brockett, R.W., 1983. Private communication.

[7] Coppel, W.A., 1978. *Dichotomies in Stability Theory*, Springer-Verlag, Berlin.

[8] Craig, J.J., 1989. *Introduction to Robotics: Mechanics and Control, Second Ed.*, Addison-Wesley, Reading, MA.

[9] Denavit, J., and Hartenberg, R.S., 1955. "A Kinematic Notation for Lower-Pair Mechanisms Based on Matrices," ASME: J. of Appl. Mech., Series E, pp. 215-221.

[10] Gottlieb, D.H., 1986. "Topology and Robots," *Proc. IEEE Int. Conf. on Robotics and Automation*, San Francisco, pp. 1689-1691.

[11] Hollerbach, J.M., 1985. "Optimum Kinematic Design for a Seven Degree of Freedom Manipulator," appearing in *Robotics Research—The 2nd International Symposium*, H. Hanafusa and H. Inoue, eds., MIT Press, Cambridge, MA, pp. 215-222.

[12] Hollerbach, J.M., and Suh, K.C., 1987. "Redundancy Resolution of Manipulators through Torque Optimization," *IEEE Journal of Robotics and Automation*, vol. RA-3, no. 4, pp. 308-316.

[13] Klein, C.A., 1985. "Use of Redundancy in the Design of Robotic Systems," *Robotics Research: The Second International Symposium*, Hanafusa and Inoue, eds., MIT Press, Cambridge, MA, pp. 207-214.

[14] Klein, C.A., and Huang, C.-H., 1983. "Review of Pseudoinverse Control for Use with Kinematically Redundant Manipulators," *IEEE Trans. on Systems, Man, and Cybernetics*, SMC-13, pp. 245-250.

[15] Loncaric, J., 1985. *Geometrical Analysis of Compliant Mechanisms in Robotics*, PhD Thesis, Division of Applied Sciences, Harvard University.

[16] Maciejewski, A.A., 1989. "Kinetic Limitations on the Use of Redundancy in Robotic Manipulators," *Proc. IEEE Int. Conf. on Robotics and Automation*, Scottsdale, Arizona, pp. 113-118

[17] Martin, D.P., 1990. "Mathematical Methods for Problems of Kinematic Redundancy in Robotics," Technical Report. Department of Aerospace and Mechanical Engineering, Boston University.

[18] Martin, D.P., Baillieul, J., and Hollerbach, J.M. 1989. "Resolution of Kinematic Redundancy Using Optimization Techniques" *IEEE Transactions on Robotics and Automation*, v. 5, no. 4, pp. 529-533.

[19] Nakamura, Y. and Hanafusa, H., 1987. "Optimal Redundancy Control of Robot Manipulators," *International Journal of Robotics Research*, vol. 6, no. 1, pp. 32-42.

[20] Sciavicco, L. and Siciliano, B., 1988. "A Solution Algorithm to the Inverse Kinematic Problem for Redundant Manipulators," *IEEE Journal of Robotics and Automation*, vol. 4, no. 4, pp. 403-410.

[21] Shamir, T., and Yomdin, Y., 1988. "Repeatability of Redundant Manipulators: Mathematical Solution of the Problem," *IEEE Transactions on Automatic Control*, AC-33, No. 11, pp. 1004-1009.

[22] Shamir, T., 1989. "On Singularities of Redundant Robot Arms." To appear.

[23] Whitney, D.E. 1969. "Resolved Motion Rate Control of Manipulators and Human Prostheses," *IEEE Trans. on Man-Machine Systems*, MMS-10, pp. 47-53.

[24] Wasow, W., 1965. *Asymptotic Expansions for Ordinary Differential Equations*, Interscience, New York. Republished by Dover, New York, 1987.

[25] Yoshikawa, T., 1984. "Manipulatability of Robotic Mechanisms," *Second International Symposium on Robotics Research*, Kyoto, Japan, August 20-23.

DEPARTMENT OF AERSOPACE/MECHANICAL ENGINEERING, BOSTON UNIVERSITY, BOSTON, MA, 02215

Proceedings of Symposia in Applied Mathematics
Volume **41**, 1990

Grasping and Manipulation using Multifingered Robot Hands

RICHARD M. MURRAY* AND S. SHANKAR SASTRY†

ABSTRACT. This paper presents a mathematical introduction to multifingered robot hands. We derive the kinematics and dynamics of a mechanical hand grasping an object and apply these results to the control of such a system. The approach presented is very general and can be used for studying other types of constrained robotic systems. Finally, we consider issues involved in planning finger motions when the fingers are allowed to roll on the surface of the object.

I. INTRODUCTION

In these notes we give the reader a feel for the mathematical problems involved in describing grasping and fine motion manipulation of objects with multifingered robot hands. Multifingered robot hands can be thought of as several robots (fingers) on a common base (palm) cooperatively manipulating an object. It is clear that positioning an object in space, namely specifying its position and orientation needs 6 degrees of freedom. However, dextrously manipulating objects requires far more degrees of freedom especially in the execution of tasks involving picking up an object, regrasping it and using the object. It is here that the study of multifingered hands is important. The study of multifingered hands has a long history not just in the context of robotics but also in the context of prosthesis.

In Chapter 2, we set down a brief discussion of the kinematics of a single rigid body, followed by a study of contacts and the kinematics of rolling. Rolling is an especially important way in which finger tips move over the surface of an object in order both to reposition and regrasp the object. In Section 2.4 we study the kinematics of a multifingered hand in terms of the kinematics of the individual fingers. Finally, we define grasp stability and the manipulability of grasps. The appendix contains a derivation of the contact equations in terms of the metric tensor and connection form of the surfaces in contact at the finger tip and object.

In Chapter 3, we develop the dynamics of multifingered hands by aggregating the dynamics of individual fingers with the dynamics of the grasped object and

---

1980 *Mathematics Subject Classification.* Primary 70Q05.
* Supported in part by an IBM Manufacturing Fellowship
† Supported in part by NSF-PYI grant DMC84-51129

the kinematic equations of contact. In Section 3.3 we describe a few different control techniques to follow a specified trajectory for the body and the grasp forces exerted on it.

In Chapter 4, we axiomatize the process of regrasping an object by rolling the finger tips on the surface of the object. We show how the problem of finding geodesics for singular or Carnot-Caratheodory metrics is useful in steering the finger tips from one grasp to another. We conclude with some open problems.

The discussion of this paper is a summary of our own work and that of others, notably those at Harvard, in the last few years in this area. Detailed references to these appear in the body of the notes.

II. Kinematics and Statics

This chapter provides a brief introduction to grasping and the notation used in this paper. We derive the basic velocity and force transformations for both fixed and rolling contacts. For a more complete discussion of the kinematics of grasping see Kerr [**5**] and Montana [**12**].

2.1. Rigid body kinematics. A rigid motion of an object is a motion which preserves distance and orientation. Every such rigid motion can be represented by a rotation followed by a translation. Letting $SO(3)$ represent the group of all proper 3×3 rotation matrices and $\mathbb{R}$ denote the real numbers, we can represent a rigid motion by the pair $(R, p) \in SO(3) \times \mathbb{R}^3$. We define $SE(3) = SO(3) \times \mathbb{R}^3$ to be the set of all rigid motions and note that $SE(3)$ is a manifold of dimension 6 as well as a group. It may be verified that $SE(3)$ is a Lie group.

The configuration of a rigid body with respect to some identity configuration is described by an element $g \in SE(3)$. g acting on a point attached to the body defines the new location of the point relative to its identity configuration. If $q \in \mathbb{R}^3$ is a point on the body relative to some base (world) reference frame, then the location of q with respect to that basis after the body undergoes a rigid motion g is

$$g(q) = Rq + p \tag{2.1}$$

where R and p are represented in the same basis as q. This action is shown pictorially in Figure 2.1. We refer to the absolute coordinates as the *world* or *base coordinates* and the coordinates of a point on the object relative to the identity configuration as the *body coordinates*.

An object trajectory is described by a time parameterized curve, $g(t) \in SE(3)$. The velocity of an object is a tangent vector at g, so $\dot{g} \in T_g SE(3)$. $\dot{g}$ also acts on points in $\mathbb{R}^3$, giving a velocity vector $\dot{g}(q) \in \mathbb{R}^3$. Since $SE(3)$ is a Lie group, we can associate each element of $T_g SE(3)$ with the Lie algebra $se(3) \approx T_e SE(3)$ where e is the identity element. An element $\xi \in se(3)$ can be represented as a skew symmetric matrix, $\mathcal{S} \in so(3)$ and a vector $v \in \mathbb{R}^3$. Furthermore, any skew

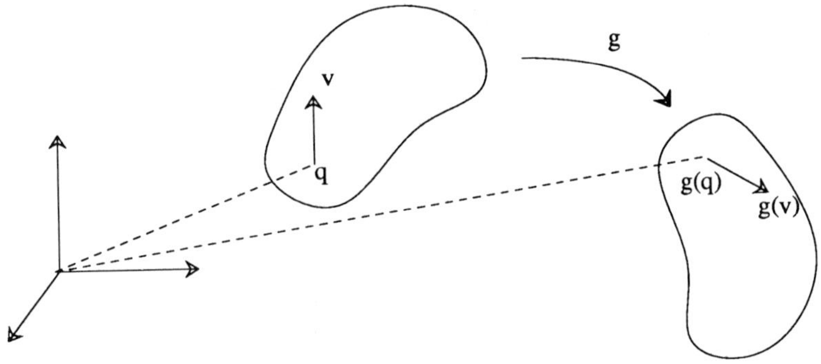

FIGURE 2.1. Rigid motion

symmetric matrix has the form:

$$\mathcal{S} = \begin{bmatrix} 0 & -\omega_z & \omega_y \\ \omega_z & 0 & -\omega_x \\ -\omega_y & \omega_x & 0 \end{bmatrix} \tag{2.2}$$

and hence we will often write $\mathcal{S}(\omega) \in so(3)$ to be the skew symmetric matrix associated with $\omega \in \mathbb{R}^3$. Note that $\mathcal{S}(\omega)q = \omega \times q$.

There are two ways to map $T_gSE(3)$ to $T_eSE(3)$ — left and right translation. The usual method is to use left translation, $L_{g^{-1}}$, where $L_gh = g \circ h$. The tangent map of $L_{g^{-1}}$ maps $T_gSE(3)$ to $T_eSE(3)$ and when applied to $\dot{g}$, the resulting map, $T_{g^{-1}}(L_{g^{-1}})\dot{g}$, takes a point in body coordinates to the velocity in body coordinates. For our purposes it is more natural to use the velocity of the point in world coordinates. This can be accomplished by using right translation and the resulting map takes a point in world coordinates to a velocity in world coordinates. Formally, we define the generalized velocity, $\xi \in T_eSE(3)$, in terms of $\dot{g} \in T_gSE(3)$ as

$$\xi = \dot{g}g^{-1} \tag{2.3}$$

The generalized velocity ξ is also called a *twist*.

Elements of $SE(3)$ can be represented as 4×4 matrices, referred to as *homogeneous coordinates*. If $g \in SE(3)$ we write

$$g = \begin{bmatrix} R & p \\ 0 & 1 \end{bmatrix} \tag{2.4}$$

A point $q \in \mathbb{R}^3$ can be represented as a vector in $\mathbb{R}^4$ by defining $\tilde{q} = (q, 1) \in \mathbb{R}^3 \times \mathbb{R}$. Using this representation, $g(q)$ becomes matrix multiplication

$$g\tilde{q} = \begin{bmatrix} R & p \\ 0 & 1 \end{bmatrix} \begin{pmatrix} q \\ 1 \end{pmatrix} = \begin{pmatrix} Rq + p \\ 1 \end{pmatrix} \tag{2.5}$$

To simplify notation we shall usually refer to $\tilde{q}$ simply as q.

The generalized velocity of a motion, in world coordinates, is

$$\xi = \dot{g}g^{-1} = \begin{bmatrix} \dot{R}R^T & \dot{p} - \dot{R}R^T p \\ 0 & 0 \end{bmatrix} \tag{2.6}$$

which can be rewritten as

$$\xi = \begin{bmatrix} \mathcal{S}(\omega) & v \\ 0 & 0 \end{bmatrix} \tag{2.7}$$

where $\omega \in \mathbb{R}^3$, $v \in \mathbb{R}^3$ and $\mathcal{S}(\omega)$ is the skew symmetric matrix generated by ω. The vector

$$\hat{\xi} = \begin{pmatrix} v \\ \omega \end{pmatrix} \tag{2.8}$$

is referred to as the *twist coordinates* of ξ and represents the rotational and linear velocity of an object as viewed in world coordinates.

2.2. Fixed contact kinematics. Traditionally, a *fixed contact* between a finger and an object is described as a mapping between forces exerted by the finger at the point of contact and the resultant forces at some reference point on the object (e.g., the center of mass). We represent the force exerted at the ith contact as $\tilde{F}_{c_i} = (\tilde{f}_{c_i}, \tilde{\tau}_{c_i}) \in \mathbb{R}^6$ where $\tilde{f}_{c_i}$ is the force exerted by contact and $\tilde{\tau}_{c_i}$ is the moment. The relationship between contact force and object force has the form

$$F_o = \begin{pmatrix} f_o \\ \tau_o \end{pmatrix} = \begin{pmatrix} \tilde{f}_{c_i} \\ \tilde{\tau}_{c_i} + r_{c_i} \times \tilde{f}_{c_i} \end{pmatrix} = \begin{bmatrix} I & 0 \\ \mathcal{S}(r_{c_i}) & I \end{bmatrix} \tilde{F}_{c_i} \tag{2.9}$$

where $r_{c_i} \in \mathbb{R}^3$ is the vector between the object reference point and the contact.

Typically, a finger will not be able to exert forces in every direction; several simple contact models are used to classify common contact configurations. A *point contact* is obtained when there is no friction between the fingertip and the object. In this case, forces can only be applied in the direction normal to the surface of the object and hence we can represent the applied force as

$$\tilde{F}_{c_i} = \begin{bmatrix} n_{c_i} \\ 0 \end{bmatrix} f_{c_i} \tag{2.10}$$

where n_{c_i} is the unit vector normal to the object and $f_{c_i} \in \mathbb{R}$ is the amount of force applied by the finger in that direction.

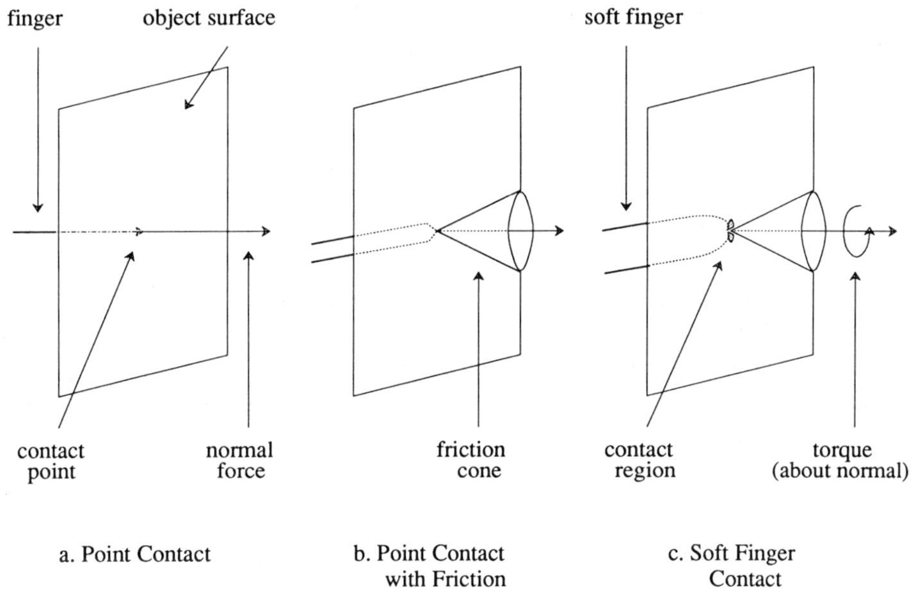

FIGURE 2.2. Contact types

A *point contact with friction* model is used when friction exists between the fingertip and the object, in which case forces can be exerted in any direction that is within a cone of forces about the direction of the surface normal. This cone, called the *friction cone*, is determined by the coefficient of friction. Figure 2.2b shows a point contact with friction and the resultant friction cone. This model assumes that moments cannot be applied (i.e., there is no torsional friction about the surface normal). As before, we represent the force felt by the object with respect to a basis of directions which are consistent with the friction model:

$$\tilde{F}_{c_i} = \begin{bmatrix} I \\ 0 \end{bmatrix} f_{c_i} \tag{2.11}$$

with $f_{c_i} \in \mathbb{R}^3$.

A more realistic contact model is the *soft finger* contact. Here we allow not only forces to be applied in a cone about the surface normal but also torques about that normal (see Figure 2.2c). These torques are limited by the torsional friction coefficient. Inside the relevant friction cones, this contact can be described as

$$\tilde{F}_{c_i} = \begin{bmatrix} I & 0 \\ 0 & n_{c_i} \end{bmatrix} \begin{pmatrix} f_{c_i} \\ \tau_{c_i} \end{pmatrix} \tag{2.12}$$

where $f_{c_i} \in \mathbb{R}^3$ and $\tau_{c_i} \in \mathbb{R}$.

Matrices mapping finger forces to contact force as in equations (2.10), (2.11) and (2.12) are referred to as *selection matrices* and we denote them by $B_i(x_o) \in \mathbb{R}^{6 \times m_i}$, where m_i is the dimension of the range of forces and moments that can be applied for a given contact type. Note their dependence on the (fixed) contact point and the orientation of the object. Each of the contact types thus can be represented as a linear map $G_i(r_{c_i}, x_o)\colon F_{c_i} \in \mathbb{R}^{m_i} \mapsto F_o$

$$G_i(r_{c_i}, x_0) = \begin{bmatrix} I & 0 \\ \mathcal{S}(r_{c_i}) & I \end{bmatrix} B_i(x_0) \qquad (2.13)$$

Since r_{c_i} is a function of the object orientation, we shall usually write $G_i(r_{c_i}, x_o)$ as $G_i(x_o)$.

If we have several fingers contacting an object then the net force on the object is the sum of the forces due to each finger. The *grasp map*, $G\colon \mathbb{R}^m \to \mathbb{R}^6$, is the map between finger forces and the resultant total object force. Since each contact map is linear and forces can be superposed, we can add the individual contact maps to form G:

$$F_o = \begin{bmatrix} G_1 & \cdots & G_k \end{bmatrix} \begin{pmatrix} F_{c_1} \\ \vdots \\ F_{c_k} \end{pmatrix} = GF_c, \qquad \begin{array}{l} F_o \in \mathbb{R}^6 \\ F_c \in \mathbb{R}^{m_1} \times \mathbb{R}^{m_2} \times \ldots \times \mathbb{R}^{m_k} \end{array} \qquad (2.14)$$

The null space of the grasp map corresponds to finger forces which cause no net force to be exerted on the object. We call the force on the object resulting from finger forces which lie in the null space of G, denoted $\mathcal{N}(G)$, *internal* or *null forces*. It is in part these internal forces which allow us to grip or squeeze an object.

Dual to the representation of contacts as applied force and torque, one may also represent a contact as a constraint between the relative velocity of the object and the finger. Letting v_{c_i} and ω_{c_i} represent the linear and angular velocity of the contact point and v_o and ω_o represent the object velocity,

$$\begin{pmatrix} \tilde{v}_{c_i} \\ \tilde{\omega}_{c_i} \end{pmatrix} = \begin{bmatrix} I & \mathcal{S}(r_{c_i}) \\ 0 & I \end{bmatrix} \begin{pmatrix} v_o \\ \omega_o \end{pmatrix} \qquad (2.15)$$

If we define v_c to be the velocities conjugate to f_c, the forces exerted by the fingers, it follows that

$$\begin{pmatrix} v_c \\ \omega_c \end{pmatrix} = G^T \begin{pmatrix} v_o \\ \omega_o \end{pmatrix} \qquad (2.16)$$

This relationship between object velocity and finger velocities can also be derived in a more general setting using the principle of virtual work.

Example. Consider a simple two-fingered planar hand as shown in Figure 2.3. Since we are in the plane, the grasp matrix maps finger forces into x and y forces, and a torque perpendicular to the xy plane. If we assume that the contacts are

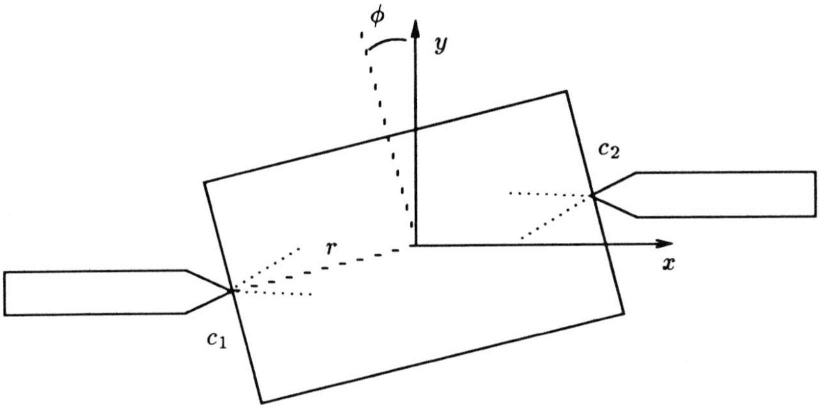

FIGURE 2.3. Planar two fingered grasp

point contacts with friction,

$$G_i(x, y, \phi) = \begin{bmatrix} I & 0 \\ \mathcal{S}\begin{pmatrix} \pm r\cos\phi \\ \pm r\sin\phi \\ 0 \end{pmatrix} & I \end{bmatrix} \begin{bmatrix} 1 & 0 \\ 0 & 1 \\ 0 & 0 \\ 0 & 0 \\ 0 & 0 \\ 0 & 0 \end{bmatrix} \tag{2.17}$$

and the planar grasp map for Figure 2.3 is

$$G(x, y, \phi) = \underbrace{\begin{bmatrix} 1 & 0 \\ 0 & 1 \\ r\sin(\phi) & -r\cos(\phi) \end{bmatrix.}_{G_1} \underbrace{\begin{matrix} 1 & 0 \\ 0 & 1 \\ -r\sin(\phi) & r\cos(\phi) \end{matrix} \end{bmatrix}}_{G_2} \tag{2.18}$$

where all forces are measured with respect to the xy coordinates shown in the figure.

Equation (2.18) shows that x and y forces from the fingers cause the same x and y forces to be exerted on the object as well as a torque that is dependent on the orientation of the object. The null space of this map is spanned by the vector

$$\begin{pmatrix} \cos\phi \\ \sin\phi \\ -\cos\phi \\ -\sin\phi \end{pmatrix} \tag{2.19}$$

which corresponds to forces applied along the line connecting the two fingertips. Finger forces applied along this line will cause no net force on the object.

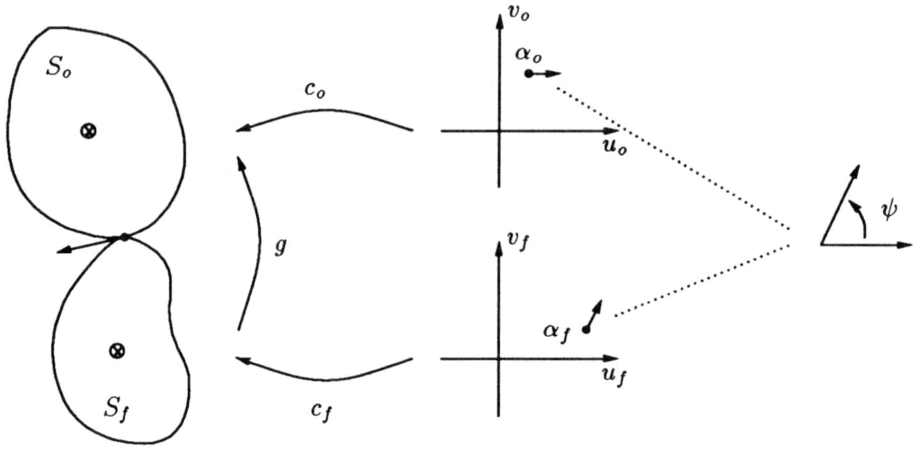

FIGURE 2.4. Parameterization of rolling contacts

2.3. Rolling contact kinematics. Most real world grasping situations involve moving rather than fixed contacts. Human fingers and many robotic fingers are actually surfaces and manipulation of an object by a set of fingers involves rolling of the fingers along the object surface. In this section we derive the kinematic equations for one object rolling against another.

Consider two objects, S_o and S_f in $\mathbb{R}^3$ which are touching at a point. We will restrict ourselves to the case where motion is contained in a single coordinate chart for each object. Let (c_o, U_o) and (c_f, U_f) be charts for the two surfaces and $\alpha_o = (u_o, v_o) \in U_o$ and $\alpha_f = (u_f, v_f) \in U_f$ be local coordinates. We will assume that c_o and c_f are orthogonal representations of the surface.[1] Furthermore, we let ψ represent the relative orientation of the tangent planes at the point of contact (see Figure 2.4). We call $\eta = (\alpha_o, \alpha_f, \psi)$ the *contact coordinates.*

Let $g \in SE(3)$ describe the relative position and orientation of S_f with respect to S_o. We wish to study the relationship between g and the local contact coordinates. To do so we assume that $g \in W \subset SE(3)$ where W is the set of all relative positions for which the two objects remain in contact.

We begin by writing the algebraic equations that η must satisfy. At any point of contact the location of the contact in space must agree for both objects

$$g \circ c_f(\alpha_f) = c_o(\alpha_o) \tag{2.20}$$

Furthermore, the tangent planes must coincide and hence the outward surface normals $n_o\colon S_o \to S^2 \subset \mathbb{R}^3$ and $n_f\colon S_f \to S^2 \subset \mathbb{R}^3$ must also agree. Letting $R \in SO(3)$ be the rotational component of g

$$Rn_f(\alpha_f) = -n_o(\alpha_o) \tag{2.21}$$

[1] A surface representation $c\colon (u,v) \to R^3$ is orthogonal if $\frac{\partial c}{\partial u}$ and $\frac{\partial c}{\partial v}$ are orthogonal. Such a representation can always be constructed for a regular surface in a given coordinate chart.

Finally, we define the angle between the tangent planes as the unique angle $\psi \in [0, 2\pi)$ such that

$$R\frac{\partial c_f}{\partial \alpha_f} M_f^{-1} R_\psi = \frac{\partial c_o}{\partial \alpha_o} M_o^{-1} \tag{2.22}$$

where

$$M_i = \begin{bmatrix} \|\frac{\partial c_i}{\partial u_i}\| & 0 \\ 0 & \|\frac{\partial c_i}{\partial v_i}\| \end{bmatrix} \tag{2.23}$$

insures that the columns of $\frac{\partial c}{\partial \alpha}$ are unit length and

$$R_\psi = \begin{bmatrix} \cos\psi & -\sin\psi \\ -\sin\psi & -\cos\psi \end{bmatrix} \tag{2.24}$$

converts α_o coordinates to the equivalent α_f coordinates at the point of contact. Since the normals are in opposite directions, R_ψ acts by negating the y coordinate and rotating by an angle ψ. Note that $R_\psi = R_\psi^T = R_\psi^{-1}$.

PROPOSITION 1. *There is a smooth local bijection between η and $g \subset W$ if and only if*

$$\frac{\partial n_o}{\partial \alpha_o} M_o^{-1} + R\frac{\partial n_f}{\partial \alpha_f} M_f^{-1} R_\psi$$

is full rank

Proof. Functionally, equations (2.20) through (2.22) are of the form $h(g, \eta) = 0$. It is therefore sufficient (and necessary) to show that $\frac{\partial h}{\partial \eta}$ spans the allowable velocity space, TW. Since ψ can be defined directly, we omit the ψ coordinate and consider the dependence on $\alpha = (\alpha_o, \alpha_f)$,

$$h(g, \alpha) = \begin{pmatrix} c_o(\alpha_o) - gc_f(\alpha_f) \\ n_o(\alpha_o) + Rn_f(\alpha_f) \end{pmatrix} \tag{2.25}$$

$$\frac{\partial h}{\partial \alpha}(g, \alpha) = \begin{pmatrix} \frac{\partial c_o(\alpha_o)}{\partial \alpha_o} & -R\frac{\partial c_f(\alpha_f)}{\partial \alpha_f} \\ \frac{\partial n_o(\alpha_o)}{\partial \alpha_o} & R\frac{\partial n_f(\alpha_f)}{\partial \alpha_f} \end{pmatrix} \tag{2.26}$$

First we show that the span of the rows of $\frac{\partial h}{\partial \alpha}$ does not contain either $(n_o, 0)$ or $(0, n_o)$, corresponding to translation and rotation about n_o. $(0, n_o)$ is spanned directly by $d\psi$ and $(n_o, 0)$ should not belong to the row span of $\frac{\partial h}{\partial \alpha}$ because motion in the n_o direction is not contained in TW. Since the range of $\frac{\partial c_o}{\partial \alpha_o}$ and $\frac{\partial c_f}{\partial \alpha_f}$ define the tangent plane and the n_i's have unit magnitude and using equation (2.21), we find

$$\begin{pmatrix} n_o \\ 0 \end{pmatrix}^T \frac{\partial h}{\partial \alpha} = \begin{bmatrix} n_o^T \frac{\partial c_o}{\partial \alpha_o} & n_f^T R^T R \frac{\partial c_f}{\partial \alpha_f} \end{bmatrix} = 0 \tag{2.27}$$

$$\begin{pmatrix} 0 \\ n_o \end{pmatrix}^T \frac{\partial h}{\partial \alpha} = \begin{bmatrix} n_o^T \frac{\partial n_o}{\partial \alpha_o} & n_f^T R^T R \frac{\partial n_f}{\partial \alpha_f} \end{bmatrix} = 0 \tag{2.28}$$

Next we examine the conditions under which $\frac{\partial h}{\partial \alpha}$ loses rank. Plugging equation (2.22) into equation (2.26), $\frac{\partial h}{\partial \alpha}$ can only lose rank when $\frac{\partial c_f}{\partial \alpha_f} = M_f^{-1} R_\psi M_o \frac{\partial c_o}{\partial \alpha_o}$, so $\frac{\partial h}{\partial \alpha}$ is full rank if and only if

$$\frac{\partial n_o}{\partial \alpha_o} M_o^{-1} + R \frac{\partial n_f}{\partial \alpha_f} M_f^{-1} R_\psi \tag{2.29}$$

is full rank. □

Proceeding along the lines of the proof given above, the differential relationship between η and g can be derived (see appendix at end of this chapter). It is convenient to make use of the *normalized gauss frame* defined on each surface

$$\begin{bmatrix} x_i & y_i & z_i \end{bmatrix} = \begin{bmatrix} \frac{\partial c_i}{\partial \alpha_i} M_i^{-1} & n_i \end{bmatrix} \tag{2.30}$$

If we do not allow the fingers to slide on the object (soft finger contacts) then the motion of the contacts, $\dot{\eta}$, as a function of the relative motion, (ω, v), is given by

$$\begin{aligned} \dot{\alpha}_o &= M_o^{-1}(K_o + \tilde{K}_f)^{-1} \omega_t \\ \dot{\alpha}_f &= M_f^{-1} R_\psi (K_o + \tilde{K}_f)^{-1} \omega_t \\ \dot{\psi} &= T_o M_o \dot{\alpha}_o + T_f M_f \dot{\alpha}_f \end{aligned} \tag{2.31}$$

where

$$\omega_t = \begin{bmatrix} x_o^T \\ y_o^T \end{bmatrix} n_o \times \omega \tag{2.32}$$

$$K_o = \begin{bmatrix} x_o^T \\ y_o^T \end{bmatrix} \frac{\partial n_o}{\partial \alpha_o} M_o^{-1} \tag{2.33}$$

$$\tilde{K}_f = R_\psi \begin{bmatrix} x_f^T \\ y_f^T \end{bmatrix} \frac{\partial n_f}{\partial \alpha_f} M_f^{-1} R_\psi \tag{2.34}$$

$$T_o = y_o^T \frac{\partial x_o}{\partial \alpha_o} M_o^{-1} \tag{2.35}$$

$$T_f = y_f^T \frac{\partial x_f}{\partial \alpha_f} M_f^{-1} \tag{2.36}$$

$(K_o + \tilde{K}_f)$ is called the *relative curvature* [**12**]. From equations (2.22) and (2.29) we see that the relative curvature is invertible precisely when $\frac{\partial h}{\partial \eta}$ is onto TW. We shall assume that all manipulation occurs in an open set on which the relative curvature is invertible.

We can now describe the kinematics for rolling contact—the relationship between the object velocities and a set of finger velocities. This situation is identical to that given for fixed contacts except that the vector r_{c_i} between the object reference frame and the ith contact point is now a function of η as well as the object orientation. But η is a continuous function of $g = x_o^{-1} x_{f_i}$ so we have

$$F_o = G(x_o, x_f) F_c \tag{2.37}$$

where $x_f = (x_{f_f}, \cdots, x_{f_k})$ is the position and orientation of the fingers and $F_c \in \mathbb{R}^{n_f} \times \cdots \times \mathbb{R}^{n_k}$ is the force exerted by the fingers at the contact point. As before, G is composed of matrices of the form

$$G_i(x_o, x_f) = \begin{bmatrix} I & 0 \\ \mathcal{S}(r_{c_i}) & I \end{bmatrix} B_i(x_o, x_f) \tag{2.38}$$

The velocity relationship can again be derived from the principle of virtual work or algebraically to determine

$$\begin{pmatrix} v_c \\ \omega_c \end{pmatrix} = G^T(x_o, x_f) \begin{pmatrix} v_o \\ \omega_o \end{pmatrix} \tag{2.39}$$

Examples. To illustrate the form of the contact equations, we consider two examples—a sphere rolling on a plane and a sphere rolling on another sphere. The local coordinates of the plane are choosen to be $c_o(u,v) = (u, v, 0)$. The sphere requires multiple coordinate charts to describe the entire surface, so we shall restrict ourself to the chart

$$c_f(u,v) = (\rho \cos u \cos v, -\rho \cos u \sin v, \rho \sin u) \tag{2.40}$$

where ρ is the radius of the sphere and $-\pi/2 < u < \pi/2$, $-\pi < v < \pi$. The curvature, torsion and metric tensors are easily calculated to be

$$\begin{aligned} K_o &= \begin{bmatrix} 0 & 0 \\ 0 & 0 \end{bmatrix} & K_f &= \begin{bmatrix} 1/\rho & 0 \\ 0 & 1/\rho \end{bmatrix} \\ M_o &= \begin{bmatrix} 1 & 0 \\ 0 & 1 \end{bmatrix} & M_f &= \begin{bmatrix} \rho & 0 \\ 0 & \rho \cos u \end{bmatrix} \\ T_o &= \begin{bmatrix} 0 & 0 \end{bmatrix} & T_f &= \begin{bmatrix} 0 & -1/\rho \tan u \end{bmatrix} \end{aligned} \tag{2.41}$$

Consider first a spherical finger of radius ρ rolling on a plane. The equations governing the evolution of the contact point are

$$\begin{aligned} \dot{u}_f &= \omega_1 \\ \dot{v}_f &= \sec u_f \, \omega_2 \\ \dot{u}_o &= \rho \cos \psi \, \omega_1 - \rho \sin \psi \, \omega_2 \\ \dot{v}_o &= -\rho \sin \psi \, \omega_1 - \rho \cos \psi \, \omega_2 \\ \dot{\psi} &= -\tan u_f \, \omega_2 \end{aligned} \tag{2.42}$$

where $\omega_t = (\omega_1, \omega_2)$. If our object is a sphere of unit radius instead of a plane,

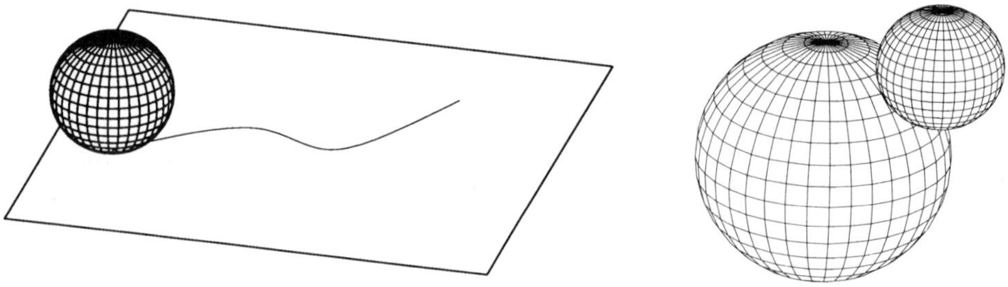

FIGURE 2.5: Spherical finger rolling on a plane and on another sphere. The finger is only allowed to roll on the object and not slip or twist.

the contact equations become

$$\begin{aligned}
\dot{u}_f &= \tfrac{1}{1+\rho}\,\omega_1 \\
\dot{v}_f &= \tfrac{1}{1+\rho}\sec u_f\,\omega_2 \\
\dot{u}_o &= \tfrac{\rho}{1+\rho}\cos\psi\,\omega_1 - \tfrac{\rho}{1+\rho}\sin\psi\,\omega_2 \\
\dot{v}_o &= -\tfrac{\rho}{1+\rho}\sin\psi\sec u_o\,\omega_1 - \tfrac{\rho}{1+\rho}\cos\psi\sec u_o\,\omega_2 \\
\dot{\psi} &= \tfrac{\rho}{1+\rho}\sin\psi\tan u_o\,\omega_1 + \tfrac{\rho}{1+\rho}(\cos\psi\tan u_o - 1rho\ \tan u_f)\,\omega_2
\end{aligned} \tag{2.43}$$

2.4. Finger Kinematics. Up to this point we have assumed that the fingers of the hand are points or surfaces in space. In fact, we are more interested in considering fingers which are kinematic mechanisms. For each finger i we associate a forward kinematic map $K_{f_i}\colon \mathbb{R}^{n_i} \to SE(3)$ which takes joint position to end effector position and orientation. The Jacobian of the forward kinematic map relates joint velocities to the end effector velocities,

$$\begin{pmatrix} v_{f_i} \\ \omega_{f_i} \end{pmatrix} = \frac{\partial K_{f_i}}{\partial \theta_{f_i}}\frac{d\theta}{dt}K_{f_i}^{-1} = J_{f_i}(\theta_{f_i})\dot{\theta}_{f_i}, \qquad \theta_{f_i} \in \mathbb{R}^{n_i},\ (v_{f_i},\omega_{f_i}) \in \mathbb{R}^6 \tag{2.44}$$

Combining this with the velocity transformation between the finger location and the contact location (a function of $x_o^{-1}x_f$) we write the *contact Jacobian* as

$$J_{c_i}(x_o,\theta_{f_i}) = \begin{pmatrix} I & \mathcal{S}(r_{c_i}) \\ 0 & I \end{pmatrix} J_{f_i} \qquad J_{c_i} : \dot{\theta} \mapsto \begin{pmatrix} v_{c_i} \\ \omega_{c_i} \end{pmatrix} \tag{2.45}$$

As with the fixed contacts, fingers are only allowed to exert forces in certain directions depending on the contact type. This is equivalent to saying that finger motions are only constrained in certain directions; these directions are given by the column span of $B_i^T(x_o,\theta_{f_i})\colon \mathbb{R}^{m_i} \to \mathbb{R}^6$ (where B_i is the selection matrix defined in Section 2.2). Combining this with the grasp map for the ith finger, we obtain the velocity constraint due to the ith contact,

$$G_i^T(x_o,\theta_{f_i})\begin{pmatrix} v_o \\ \omega_o \end{pmatrix} = B_i^T(x_o,\theta_{f_i})J_{c_i}(x_o,\theta_{f_i})\dot{\theta}_{f_i} \tag{2.46}$$

We now stack these matrices and write the grasp constraint for the hand as

$$\begin{bmatrix} G_1^T \\ \vdots \\ G_k^T \end{bmatrix} \begin{pmatrix} v_o \\ \omega_o \end{pmatrix} = \begin{bmatrix} B_1^T J_{c_1} & & 0 \\ & \ddots & \\ 0 & & B_k^T J_{c_k} \end{bmatrix} \dot{\theta} \tag{2.47}$$

$$G^T(x_o, \theta) \begin{pmatrix} v_o \\ \omega_o \end{pmatrix} = J(x_o, \theta)\dot{\theta} \tag{2.48}$$

2.5. Grasp stability and manipulability. For contact models involving friction, we must insure that all contact forces lie within the friction cone determined by the coefficient of friction. The set of all forces lying in or on the friction cone is

$$FC = \left\{ f_c \in \mathbb{R}^n : \|f_{c_{ij}}^t\| \leq \mu_{ij} \|f_{c_i}^n\|, \quad i = 1, \ldots, k, \quad j = 1, \ldots, m_i \right\} \tag{2.49}$$

where $f_{c_{ij}}^t$ is the tangent component of the jth element of f_{c_i}, $f_{c_i}^n$ is the normal force for the ith contact, and μ_{ij} is the coefficient of friction corresponding to $f_{c_{ij}}$. For soft finger contacts, the torques exerted by the fingers also satisfy equation (2.49) with $f_{c_{ij}}^t$ replaced by the torque (i.e., we do not want to apply a torque which is greater than the torsional friction coefficient multiplied by the magnitude of the normal force).

We say a grasp on an object is *stable* if we can resist, through a set of contacts, arbitrary forces and torques on the object. This requires that the image of the grasp map over the set of forces in the friction cone span the space of forces and torques on the object, that is $G(FC) = \mathbb{R}^6$. Note that this is a condition only on the contact kinematics and not the finger kinematics. A stable grasp is also called a *force closure* grasp.

A grasp is said to be *manipulable* if arbitrary motions of the object can be accommodated by the fingers. Unlike stability, manipulability is a property of both the contact and finger kinematics. Since the range of motion of the contacts is given locally by the range of the hand Jacobian, the condition can be written as $\mathcal{R}(G^T) \subseteq \mathcal{R}(J)$.

It is also useful to define the concept of prehensility. A grasp is *prehensile* if there exists a force contained in the null space of the grasp map which also lies in the *interior* of the friction cone. More formally, $\mathcal{N}(G) \cap \overset{\circ}{FC} \neq \{\}$ where $\overset{\circ}{FC}$ is the set of forces lying completely within the friction cone (i.e., $\|f_{c_{ij}}^t\| < \mu_{ij}\|f_{c_i}^n\|$). We shall require this property in order to insure that our controllers can maintain a grip on an object while manipulating it.

We shall generally assume that a grasp has been chosen which is stable, manipulable and prehensile. The problem of finding such grasps given a set of fingers and an object has been studied in some detail. A good treatment is given by Nguyen [**14**].

Appendix – Contact kinematics derivation. In this appendix we derive the kinematics of contact for two objects touching each other at a point. The notation is described more fully in Section 2.3. An alternate derivation can be found in a recent paper by Montana [**12**].

To derive the kinematics, we begin with constraint equations given by equating the points of contact, normals of contact and tangent planes at the contact points:

$$Rc_f(\alpha_f) + p = c_o(\alpha_o) \tag{2.50}$$

$$Rn_f(\alpha_f) = -n_o(\alpha_o) \tag{2.51}$$

$$R\frac{\partial c_f}{\partial \alpha_f} M_f^{-1} R_\psi = \frac{\partial c_o}{\partial \alpha_o} M_o^{-1} \tag{2.52}$$

Differentiate (2.50) and (2.51)

$$\dot{R}c_f + R\frac{\partial c_f}{\partial \alpha_f}\dot{\alpha}_f + \dot{p} = \frac{\partial c_o}{\partial \alpha_o}\dot{\alpha}_o \tag{2.53}$$

$$\dot{R}n_f + R\frac{\partial n_f}{\partial \alpha_f}\dot{\alpha}_f = -\frac{\partial n_o}{\partial \alpha_o}\dot{\alpha}_o \tag{2.54}$$

Multiply (2.53) by $\frac{\partial c_o}{\partial \alpha_o}^T$ and substitute $\dot{\alpha}_o$ into (2.54)

$$\dot{R}n_f + R\frac{\partial n_f}{\partial \alpha_f}\dot{\alpha}_f = -\frac{\partial n_o}{\partial \alpha_o} M_o^{-2} \frac{\partial c_o}{\partial \alpha_o}^T \left(\dot{R}c_f + R\frac{\partial c_f}{\partial \alpha_f}\dot{\alpha}_f + \dot{p}\right) \tag{2.55}$$

Using (2.52) in the last term of (2.55) and rearranging

$$\begin{aligned} &\left(R\frac{\partial n_f}{\partial \alpha_f} + \frac{\partial n_o}{\partial \alpha_o} M_o^{-2} (\frac{\partial c_o}{\partial \alpha_o}^T \frac{\partial c_o}{\partial \alpha_o}) M_o^{-1} R_\psi M_f\right)\dot{\alpha}_f \\ &= -\dot{R}n_f - \frac{\partial n_o}{\partial \alpha_o} M_o^{-2} \frac{\partial c_o}{\partial \alpha_o}^T (\dot{R}c_f + \dot{p}) \end{aligned} \tag{2.56}$$

Simplify the first term and multiply by $M_o^{-T}\frac{\partial c_o}{\partial \alpha_o}^T$ on the left

$$\begin{aligned} &M_o^{-T}\frac{\partial c_o}{\partial \alpha_o}^T \left(R\frac{\partial n_f}{\partial \alpha_f} + \frac{\partial n_o}{\partial \alpha_o} M_o^{-1} R_\psi M_f\right) \\ &= M_o^{-T}\frac{\partial c_o}{\partial \alpha_o}^T \left(R\frac{\partial n_f}{\partial \alpha_f} M_f^{-1} R_\psi + \frac{\partial n_o}{\partial \alpha_o} M_o^{-1}\right) R_\psi M_f \\ &= \Big(\underbrace{R_\psi M_f^{-T}\frac{\partial c_f}{\partial \alpha_f}^T \frac{\partial n_f}{\partial \alpha_f} M_f^{-1} R_\psi}_{\tilde{K}_f} + \underbrace{M_o^{-T}\frac{\partial c_o}{\partial \alpha_o}^T \frac{\partial n_o}{\partial \alpha_o} M_o^{-1}}_{K_o}\Big) R_\psi M_f \end{aligned} \tag{2.57}$$

Multiply both sides of (2.56) by $M_o^{-T}\frac{\partial c_o}{\partial \alpha_o}^T$ and use the previous calculation

$$\dot{\alpha}_f = M_f^{-1} R_\psi \left(\tilde{K}_f + K_o\right)^{-1}.$$

$$
\begin{aligned}
& M_o^{-T}\frac{\partial c_o}{\partial \alpha_o}^T\left(-\dot{R}n_f - \frac{\partial n_o}{\partial \alpha_o}M_o^{-2}\frac{\partial c_o}{\partial \alpha_o}^T(\dot{R}c_f+\dot{p})\right) \\
= \;& M_f^{-1}R_\psi\left(\tilde{K}_f+K_o\right)^{-1}\cdot \\
& \left(-M_o^{-T}\frac{\partial c_o}{\partial \alpha_o}^T\dot{R}n_f - K_oM_o^{-T}\frac{\partial c_o}{\partial \alpha_o}^T(\dot{R}c_f+\dot{p})\right)
\end{aligned}
\tag{2.58}
$$

Let ω_t stand for the $\dot{R}n_f$ term and v_t represent the $\dot{R}c_f+\dot{p}$ term:

$$
\dot{\alpha}_f = M_f^{-1}R_\psi\left(\tilde{K}_f+K_o\right)^{-1}(w_t - K_o v_t) \tag{2.59}
$$

Now w_t and v_t can now be calculated in terms of the relative velocity given by $(\mathcal{S}(\omega), v) = \dot{g}g^{-1}$. We use the fact that $\mathcal{S}(\omega)a = \omega \times a$ and $\omega \times a = -a \times \omega$ to obtain

$$
w_t = -M_o^{-T}\frac{\partial c_o}{\partial \alpha_o}^T(\omega \times (Rn_f)) = -M_o^{-T}\frac{\partial c_o}{\partial \alpha_o}^T(n_o \times \omega) \tag{2.60}
$$

$$
\begin{aligned}
v_t &= M_o^{-T}\frac{\partial c_o}{\partial \alpha_o}^T(\omega \times (Rc_f) + \omega \times p + v) \\
&= M_o^{-T}\frac{\partial c_o}{\partial \alpha_o}^T(\omega \times (c_o - p) + \omega \times p + v) \\
&= M_o^{-T}\frac{\partial c_o}{\partial \alpha_o}^T(-c_o \times \omega + v)
\end{aligned}
\tag{2.61}
$$

We see that ω_t is the relative rotational velocity projected onto the tangent plane at the contact. It includes only terms due to *rolling* since rotation normal to the surface is annihilated by taking the cross product with n_o. Likewise, v_t is the relative linear velocity between the contacts, projected onto the tangent plane, i.e., the *sliding* velocity.

A similar calculation yields

$$
\dot{\alpha}_o = M_o^{-1}\left(\tilde{K}_f+K_o\right)^{-1}\left(w_t - \tilde{K}_f v_t\right) \tag{2.62}
$$

which gives the kinematics for the object contact point in local coordinates.

Next we solve for ψ, the angle between the tangent planes of the finger and object. Combining (2.51) and (2.52) we can write

$$
R\begin{bmatrix} \frac{\partial c_f}{\partial \alpha_f}M_f^{-1} & n_f \end{bmatrix}\begin{bmatrix} R_\psi & 0 \\ 0 & -1 \end{bmatrix} = \begin{bmatrix} \frac{\partial c_o}{\partial \alpha_o}M_o^{-1} & n_o \end{bmatrix} \tag{2.63}
$$

and using the normalized gaussian coordinates this can be rewritten

$$
R[x_f\; y_f\; z_f]\bar{R}_\psi = [x_o\; y_o\; z_o] \tag{2.64}
$$

Take the derivative of (2.64)

$$
\dot{R}[x_f\, y_f\, z_f]\bar{R}_\psi + R[\dot{x}_f\, \dot{y}_f\, \dot{z}_f]\bar{R}_\psi + R[x_f\, y_f\, z_f]\begin{bmatrix} \dot{R}_\psi & 0 \\ 0 & 0 \end{bmatrix} = [\dot{x}_o\, \dot{y}_o\, \dot{z}_o] \tag{2.65}
$$

Premultiply by $y_f^T R^T$

$$y_f^T R^T \dot{R} [x_f\, y_f\, z_f] \bar{R}_\psi + y_f^T [\dot{x}_f\, \dot{y}_f\, \dot{z}_f] \bar{R}_\psi + (0\,1\,0) \begin{bmatrix} \dot{R}_\psi 0 \\ 0\; 0 \end{bmatrix} = y_f^T R^T [\dot{x}_o\, \dot{y}_o\, \dot{z}_o] \tag{2.66}$$

Postmultiply by $\bar{R}_\psi \begin{pmatrix} 1 \\ 0 \\ 0 \end{pmatrix}$ and note $\bar{R}_\psi \bar{R}_\psi = I$

$$y_f^T R^T \dot{R} x_f + y_f^T \dot{x}_f + (0\,1\,0) \begin{bmatrix} \dot{R}_\psi R_\psi & 0 \\ 0 & 0 \end{bmatrix} \begin{pmatrix} 1 \\ 0 \\ 0 \end{pmatrix} = y_f^T R^T [\dot{x}_o\, \dot{y}_o\, \dot{z}_o] \bar{R}_\psi \begin{pmatrix} 1 \\ 0 \\ 0 \end{pmatrix} \tag{2.67}$$

$$y_f^T R^T \dot{R} x_f + y_f^T \dot{x}_f + (0\,1) \begin{bmatrix} 0 & \dot{\psi} \\ -\dot{\psi} & 0 \end{bmatrix} \begin{pmatrix} 1 \\ 0 \end{pmatrix} = y_f^T R^T [\dot{x}_o\, \dot{y}_o\, \dot{z}_o] \bar{R}_\psi \begin{pmatrix} 1 \\ 0 \\ 0 \end{pmatrix} \tag{2.68}$$

$$y_f^T R^T \dot{R} x_f + y_f^T \dot{x}_f - \dot{\psi} = y_f^T R^T [\dot{x}_o\, \dot{y}_o\, \dot{z}_o] \bar{R}_\psi \begin{pmatrix} 1 \\ 0 \\ 0 \end{pmatrix} \tag{2.69}$$

From (2.64) we see that $y_f^T R^T = (0\,1\,0) \bar{R}_\psi \begin{bmatrix} x_o^T \\ y_o^T \\ z_o^T \end{bmatrix} = (0\,1) R_\psi \begin{bmatrix} x_o^T \\ y_o^T \end{bmatrix}$ and so

$$\dot{\psi} = y_f^T R^T \dot{R} x_f + y_f^T \frac{\partial x_f}{\partial \alpha_f} \dot{\alpha}_f - (0\,1) R_\psi \begin{bmatrix} x_o^T \dot{x}_o & x_o^T \dot{y}_o \\ y_o^T \dot{x}_o & y_o^T \dot{y}_o \end{bmatrix} R_\psi \begin{pmatrix} 1 \\ 0 \end{pmatrix} \tag{2.70}$$

Using the following identities

$$\begin{aligned} x_i^T y_i = 0 \quad &\Rightarrow \quad \dot{x}_i^T y_i = -x_i^T \dot{y}_i = y_i^T \dot{x}_i \\ x_i^T x_i = 1 \quad &\Rightarrow \quad \dot{x}_i^T x_i = 0 \end{aligned} \tag{2.71}$$

(2.70) can be written as

$$\begin{aligned} \dot{\psi} &= y_f^T R^T \dot{R} x_f + y_o^T \frac{\partial x_o}{\partial \alpha_o} \dot{\alpha}_o + y_f^T \frac{\partial x_f}{\partial \alpha_f} \dot{\alpha}_f \\ &= \omega_n + T_o M_o \dot{\alpha}_o + T_f M_f \dot{\alpha}_f \end{aligned} \tag{2.72}$$

where

$$\begin{aligned} \omega_n &= y_f^T R^T \dot{R} x_f = (R y_f)^T \omega \times (R x_f) \\ &= (R z_f)^T \omega = z_o^T \omega \end{aligned} \tag{2.73}$$

and the last equality follows from the vector formula $a \cdot b \times c = b \cdot c \times a$. This last equation shows that ω_n is just the relative rotational velocity projected onto the surface normal.

Collecting equations (2.59), (2.62) and (2.72) we have

$$\dot{\alpha}_o = M_o^{-1} (K_o + \tilde{K}_f)^{-1} \left(\omega_t - \tilde{K}_f v_t \right) \tag{2.74}$$

$$\dot{\alpha}_f = M_f^{-1} R_\psi (K_o + \tilde{K}_f)^{-1} (\omega_t - K_o v_t) \tag{2.75}$$

$$\dot{\psi} = \omega_n + T_o M_o \dot{\alpha}_o + T_f M_f \dot{\alpha}_f \tag{2.76}$$

The matrix $K_o + \tilde{K}_f$ is called the *relative curvature* by Montana[**12**].

III. Dynamics and Control

In this section we review some basic results in dynamics of robot systems. The primary result which we present is that even for relatively complicated robot systems, the equations of motion for the system can be written in a standard form. This point of view has been used by Khatib in his operational space formulation [**6**] and in some recent extensions [**7**]. The results presented in this section are direct extensions of those works, although the approach is different.

3.1. Robot dynamics. We begin by deriving the robot dynamics for a manipulator in joint space. Let $\theta \in \mathbb{R}^n$ be the joint angles for the manipulator and $\tau \in \mathbb{R}^n$ be the corresponding joint torques. The Lagrangian for the system may be shown to be of the form

$$L = \frac{1}{2}M(\theta)(\dot{\theta},\dot{\theta}) + V(\theta) \tag{3.1}$$

where $M(\theta)$ is the inertia matrix for the manipulator and $V(\theta)$ is the potential energy due to gravity. Substituting into Lagrange's equations

$$\left(\frac{d}{dt}\frac{\partial L}{\partial \dot{\theta}} - \frac{\partial L}{\partial \theta} - \tau\right)\delta\theta = 0 \tag{3.2}$$

and letting τ represent the actuator torques (and other non-conservative forces), we obtain

$$M(\theta)(\ddot{\theta},\cdot) + DM(\theta)(\dot{\theta},\cdot)(\dot{\theta}) - \frac{1}{2}DM(\theta)(\dot{\theta},\dot{\theta})(\cdot) + DV(\theta)(\cdot) = \tau \tag{3.3}$$

To put this in a more conventional form we define the matrix $C(\theta,\dot{\theta})$ as

$$a^T C(\theta,\dot{\theta})b = \frac{1}{2}DM(\theta)(\dot{\theta},a)(b) + \frac{1}{2}DM(\theta)(b,a)(\dot{\theta}) - \frac{1}{2}DM(\theta)(\dot{\theta},b)(a) \tag{3.4}$$

and write

$$M(\theta)\ddot{\theta} + C(\theta,\dot{\theta})\dot{\theta} + N(\theta,\dot{\theta}) = \tau \tag{3.5}$$

where $N(\theta,\dot{\theta})$ includes gravity terms and other forces (such as friction) which act at the joints.

For systems of this type, the inertia matrix is always symmetric and positive definite and it can be shown that $\dot{M} - 2C$ is skew symmetric (using this particular choice of C). It is both the form and the structure of this equation that we will attempt to maintain in more complicated systems.

3.2. Robot hand dynamics. We now examine the dynamics of a set of fingers actuated at each joint connected through a set of contacts to a rigid body. The finger dynamics can be written as

$$M_f(\theta)\ddot{\theta} + C_f(\theta,\dot{\theta})\dot{\theta} + N_f(\theta,\dot{\theta}) = \tau \tag{3.6}$$

where $\theta \in \mathbb{R}^{n_1} \times \cdots \times \mathbb{R}^{n_k}$ is now the set of joint angles for *all* of the robots and τ is the corresponding set of torques. The object dynamics are given by the Newton-Euler equations

$$\begin{bmatrix} m_o I & 0 \\ 0 & \mathcal{I}_o \end{bmatrix} \begin{pmatrix} \dot{v}_o \\ \dot{\omega}_o \end{pmatrix} + \begin{pmatrix} 0 \\ \omega_o \times \mathcal{I}_o \omega_o \end{pmatrix} + \begin{pmatrix} DV_o(x_o) \\ 0 \end{pmatrix} = \begin{pmatrix} f_o \\ \tau_o \end{pmatrix} \tag{3.7}$$

where $\mathcal{I}_o = R\mathcal{I}R^T$ is the object inertia in world coordinates and V_o is the potential energy. In local coordinates this has the same basic form as the robot dynamics, lacking only the actuator torques:

$$M_o(x)\ddot{x} + C_o(x, \dot{x})\dot{x} + N_o(x, \dot{x}) = 0 \tag{3.8}$$

where x is a local parameterization of $x_o \in SE(3)$. We attach these two systems with a set of constraints

$$G^T(x, \theta)\dot{x} = J(x, \theta)\dot{\theta} \tag{3.9}$$

which represents the grasp. We will assume that the grasp is both stable and manipulable. For the moment we will also require J to be injective.

This velocity constraint generates a constraint on the virtual displacements $\delta\theta$ and δx, namely $\delta\theta = J^{-1}(q)G^T(q)\delta x$ with $q = (x, \theta)$. Using this relationship, we can write Lagrange's equations as

$$\left(\frac{d}{dt}\frac{\partial L}{\partial q} - \frac{\partial L}{\partial q} - (\tau, 0)\right)\delta q = 0 \tag{3.10}$$

$$\begin{pmatrix} \frac{d}{dt}\frac{\partial L}{\partial \dot{\theta}} - \frac{\partial L}{\partial \theta} - \tau \\ \frac{d}{dt}\frac{\partial L}{\partial \dot{x}} - \frac{\partial L}{\partial x} \end{pmatrix} \begin{pmatrix} \delta\theta \\ \delta x \end{pmatrix} = 0 \tag{3.11}$$

$$\left(\frac{d}{dt}\frac{\partial L}{\partial \dot{\theta}} - \frac{\partial L}{\partial \theta} - \tau\right)\delta\theta + \left(\frac{d}{dt}\frac{\partial L}{\partial \dot{x}} - \frac{\partial L}{\partial x}\right)\delta x = 0 \tag{3.12}$$

$$GJ^{-T}\left(\frac{d}{dt}\frac{\partial L}{\partial \dot{\theta}} - \frac{\partial L}{\partial \theta} - \tau\right)\delta x + \left(\frac{d}{dt}\frac{\partial L}{\partial \dot{x}} - \frac{\partial L}{\partial x}\right)\delta x = 0 \tag{3.13}$$

and since δx is arbitrary

$$\frac{d}{dt}\frac{\partial L}{\partial \dot{x}} - \frac{\partial L}{\partial x} + GJ^{-T}\left(\frac{d}{dt}\frac{\partial L}{\partial \dot{\theta}} - \frac{\partial L}{\partial \theta}\right) = GJ^{-T}\tau \tag{3.14}$$

This equation together with the velocity constraint given in equation (3.9) describes the system completely. Note that equation (3.14) is a vector differential equation with $n - m$ rows and equation (3.9) is a vector equation with m rows.

It is tempting to derive equation (3.14) by using the velocity constraint directly in the kinetic energy equation (which is a function of $\dot{\theta}$ and $\dot{x}$) and then substituting this into Lagrange's equations. As noted in Rosenberg [**15**] this can only be done if the constraint is holonomic, i.e., θ can be written as a function of x.

Next we separate the kinetic energy into an object portion and a robot portion

$$T = \dot\theta^T M_f(\theta)\dot\theta + \dot x^T M_o(x)\dot x \tag{3.15}$$

Using equation (3.14) we find

$$\tilde M(q)\ddot x + \tilde C(q,\dot q)\dot x \;=\; GJ^{-T}\tau = F \tag{3.16}$$

where

$$\begin{aligned}\tilde M &= M_o + GJ^{-T}M_fJ^{-1}G^T\\ \tilde C &= C_o + GJ^{-T}\left(C_fJ^{-1}G^T + M_f\frac{d}{dt}\left(J^{-1}G^T\right)\right)\end{aligned}$$

and C_f and C_o are obtained from equation (3.4) by replacing M with M_f and M_o respectively. It can be shown that the matrix $\dot{\tilde M} - 2\tilde C$ is still skew symmetric.

Thus we have an equation with form (and structure) similar to our "simple" robot. In the object frame of reference, $\tilde M$ is the effective mass of the object, and $\tilde C$ is the effective Coriolis and centrifugal matrix. These matrices include the dynamics of the fingers, which are being used to actually control the motion of the object. However the details of the finger kinematics and dynamics are effectively hidden in the definition of $\tilde M$ and $\tilde C$.

This simple result has some interesting consequences in control. Typically robot controllers are designed by placing a feedback loop around the joint positions (and velocities) of the robot. The controller generates torques which attempt to make the robot follow a prescribed joint trajectory. This can lead to difficulty in grasping situations since the joint level controllers are often not aware of the constraints and therefore may violate them. However, since the grasping dynamics are of the same form as the dynamics of a single manipulator, we can just as easily write the control algorithm in object coordinates. An additional advantage of this approach is that controller objectives are often specified in terms of the object motion and hence it might be easier to perform the controller design and analysis in that space.

Even though we will write our controllers in terms of F, it is actually the joint torques which we are able to specify. Given the desired force in constrained coordinates, we can apply that force using an actuator force of $J^TG^+\tau$, where G^+ is a pseudo-inverse for G. In general G is not square and by examining the right side of the equations of motion (3.16) we note that if $J^{-T}\tau \in \mathcal{N}(G)$ then the net force in the object frame of reference is zero and hence forces of this form cause no net motion on the object. These forces are in fact the forces which act against the constraint and are generally termed *internal* or *constraint* forces. We can use these internal forces to satisfy other conditions, such as keeping the contact forces inside the friction cone (to avoid slipping) or varying the load distribution of a set of manipulators rigidly grasping an object.

Redundant manipulators. Some manipulators contain more degrees of freedom than are necessary to specify the position of the end effector. Mathematically, these robots can be represented by a change of coordinates $f : \mathbb{R}^m \to \mathbb{R}^n$ where $m > n$. In this case $J := \frac{\partial f}{\partial \theta}$ is not square and hence J^{-1} is not well defined so our derivation of equation (3.16) does not hold.

It is still possible to write the dynamics of redundant manipulators in a form consistent with equation (3.16). To do so, we first define a matrix $K(\theta)$ whose rows span the null space of $J(\theta)$. As before we assume that $J(\theta)$ is full row rank and hence $K(\theta)$ has constant rank $m - n$. The rows of $K(\theta)$ are basis elements for the space of velocities which cause no motion of the end effector; we can thus define an *internal motion*, $\dot{y} \in \mathbb{R}^{m-n}$ using the equation

$$\begin{pmatrix} \dot{x} \\ \dot{y} \end{pmatrix} = \begin{bmatrix} J \\ K \end{bmatrix} \dot{\theta} \tag{3.17}$$

and our constraint becomes

$$\bar{J}\dot{\theta} = \begin{bmatrix} J \\ K \end{bmatrix} \dot{\theta} = \begin{bmatrix} G^T & 0 \\ 0 & I \end{bmatrix} \begin{bmatrix} \dot{x} \\ \dot{y} \end{bmatrix} = \bar{G} \begin{pmatrix} \dot{x} \\ \dot{y} \end{pmatrix} \tag{3.18}$$

The kinetic energy can be written

$$T = \dot{\theta}^T M_f(\theta)\dot{\theta} + \begin{pmatrix} \dot{x} \\ \dot{y} \end{pmatrix}^T \begin{bmatrix} M_o & 0 \\ 0 & 0 \end{bmatrix} \begin{pmatrix} \dot{x} \\ \dot{y} \end{pmatrix} \tag{3.19}$$

and since $\bar{J}$ is invertible it follows from our previous derivation that

$$\tilde{M}(q) \begin{pmatrix} \ddot{x} \\ \ddot{y} \end{pmatrix} + \tilde{C}(q, \dot{q}) \begin{pmatrix} \dot{x} \\ \dot{y} \end{pmatrix} + \tilde{N}(q, \dot{q}) = \bar{G}\bar{J}^{-T}\tau \tag{3.20}$$

where $\tilde{M}$ and $\tilde{C}$ are obtained as in the nonredundant case but replacing J with $\bar{J}$ and G with $\bar{G}$. If we choose K such that its rows are orthonormal then $\bar{J}^{-1} = (J^+ \ K^T)$ where $J^+ = J^T(JJ^T)^{-1}$ is the least-squares right (pseudo) inverse of J.

As before, the structure of the robot dynamics is maintained. However, the inertia and Coriolis/centrifugal matrices are functions of the robot configuration, θ, and not the end effector configuration, x. This complication is unavoidable due to the non-uniqueness of the inverse kinematic problem. In principle, one could locally parameterize the redundant motion by y (the integral of $\dot{y}$), allowing these matrices to be written as functions of both x and y; we will not assume that such a parameterization is available.

3.3. Control. The grasping control problem can be broken into two parts

1. Tracking – the center of mass of the object should follow a specified trajectory.

2. Holding – the finger forces should lie within the friction cone at all times.

Condition 2 is important not only because we do not wish to lose our grip on the object, but also because we assumed in our derivation of the grasp dynamics that contact was maintained. Without this constraint we would have to specify the dynamics of contact.

If a grasp is prehensile it can be shown that given an arbitrary set of finger forces, F_c, we can find an internal force, $F_N \in \mathcal{N}(G)$, such that the combined force $F_c + F_N$ is inside the friction cone. Thus, given a force generated to solve the tracking problem, we can always add a force to this such that condition 2 is satisfied. Since internal forces cause no net motion of the hand or object, this additional force does not affect the net force exerted by the fingers on the object. We shall assume in the sequel that such an internal force is available at all times. The choice of this force is discussed in more detail below.

To illustrate the control of robot systems, we look at two controllers which have appeared in the robotics literature. We consider only grasps which are stable, manipulable and prehensile. We start by considering systems of the form

$$M(q)\ddot{x} + C(q,\dot{q})\dot{x} + N(q,\dot{q}) = F \tag{3.21}$$

where $M(q)$ is a positive definite inertia matrix and $C(q,\dot{q})\dot{x}$ is the Coriolis and centrifugal force vector. The vector $N(q,\dot{q}) \in \mathbb{R}^n$ contains all friction and gravity terms and the vector $F \in \mathbb{R}^n$ represents generalized forces in the object coordinate frame. Given an object force F, we apply that force by commanding a set of joint torques

$$\tau = J^T G^+ F + J^T F_N \tag{3.22}$$

where J and G define the grasping constraint and $F_N \in \mathcal{N}(G)$.

Computed torque. Computed torque is an exactly linearizing control law (i.e., the dynamics are rendered linear by state feedback) that has been used extensively in robotics research. It has been used for joint level control [**1**], Cartesian control [**11**], and most recently, control of multi-fingered hands [**10, 4**]. Given a desired trajectory x_d we use the control

$$F = M(q)\left(\ddot{x}_d + K_v\dot{e} + K_p e\right) + C(q,\dot{q})\dot{x} + N(q,\dot{q}) \tag{3.23}$$

where error $e = x_d - x$ and K_v and K_p are constant gain matrices. The resulting dynamics equations are linear with exponential rate of convergence determined by K_v and K_p. Since the system is linear, we can use linear control theory to choose the gains (K_v and K_p) such that they satisfy some set of design criteria.

The disadvantage of this control law is that it is not easy to specify the interaction with the environment. From the form of the error equation we might think that we could use K_p to model the stiffness of the system and exert forces by commanding trajectories which result in fixed errors. Unfortunately this is not uniformly applicable as can be seen by examining the force due to a quasi-static displacement Δx:

$$\Delta F = M(q)K_p\Delta x \tag{3.24}$$

Since K_p must be constant in order to prove stability, the resultant stiffness will vary with configuration. Additionally, given a desired stiffness matrix it may not be possible to find a positive definite K_p that achieves that stiffness.

'PD' control. PD controllers differ from computed torque controllers in that the desired stiffness (and potentially damping) of the end effector is specified, rather than its position tracking characteristics. Typically, control laws of this form rely on the skew symmetric property of robot dynamics, namely $\alpha^T \left(\dot{M} - 2C\right) \alpha = 0$ for all $\alpha \in \mathbb{R}^n$, for proof of stability. Consider the control law

$$F = M(q)\ddot{x}_d + C(q,\dot{q})\dot{x}_d + N(q,\dot{q}) + K_v\dot{e} + K_p e \tag{3.25}$$

where K_v and K_p are symmetric positive definite. Using a Liapunov stability argument, it can be shown that the actual trajectory of the robot converges to the desired trajectory asymptotically [**8**]. Extensions to the control law result in exponential rate of convergence [**17, 16**].

This PD control law has the advantage that for a quasi-static change in position Δx the resulting force is

$$\Delta F = K_p \Delta x \tag{3.26}$$

and thus we can achieve an arbitrary symmetric stiffness. Experimental results indicate that the trajectory tracking performance of this control law does not always compare favorably with the computed torque control law [**13**]. Additionally there is no simple design criteria for choosing K_v and K_p to achieve good tracking performance. While the stability results give necessary conditions for stability they do not provide a method for choosing the gains. Nonetheless, PD control has been used effectively in many robot controllers and has some computational features which make it an attractive alternative.

Internal forces. All of the controllers rely on the choice of a grasping force, $F_N \in \mathcal{N}(G)$ which maintains contact between the fingertips and the object by insuring that the finger forces lie in the friction cone. There are several possible methods for calculating this term. Since F_N does not affect the motion of the object, its choice does not affect object tracking. We begin by showing that given any desired object force, there exists a set of finger forces lying in the friction cone which achieves it.

PROPOSITION 2. *If a grasp is prehensile, given any $F_c \in \mathbb{R}^m$, there exists a null force, F_N, such that the total finger force, $F_c + F_N$, is inside the friction cone.*

Proof. By the definition of a prehensile grasp there exists $F_N \in \mathcal{N}(G) \cap \overset{\circ}{FC}$ such that

$$\|F^t_{N_{ij}}\| < \mu_i \|F^n_{N_i}\| \tag{3.27}$$

where $F^t_{N_{ij}}$ is the tangent component of F_N projected onto the jth force direction of the ith contact and F_{N_i} is the normal component of F_N at the ith contact point F_{N_i} is nonzero for each i and therefore by increasing F_N, we always increase the normal component of the force exerted at each contact with respect to the tangential forces. Since $\overset{\circ}{FC}$ is defined as the Cartesian product of the n friction cones in equation (3.27), $F_N \in \mathcal{N}(G) \cap \overset{\circ}{FC}$ implies $\alpha F_N \in \mathcal{N}(G) \cap \overset{\circ}{FC}$ for all $\alpha \in \mathbb{R}^+$. Now we can look at the unit vector in the $F_c + \alpha F_N$ direction as $\alpha \to \infty$:

$$\lim_{\alpha \to \infty} \frac{F_c + \alpha F_N}{\|F_c + \alpha F_N\|} = F_N \tag{3.28}$$

Since $F_N \in \overset{\circ}{FC}$ it follows that for sufficiently high α, $\frac{F_c + \alpha F_N}{\|F_c + \alpha F_N\|}$ is also in $\overset{\circ}{FC}$ and hence $F_c + \alpha F_N$ is in the interior of the FC. Now from the definition of FC, the individual contact forces must all lie within their respective friction cones simultaneously. □

The simplest F_N is a constant F_N. It must be large enough so that finger forces never leave the friction cone over the entire trajectory of the object. Generally this requires a knowledge of the bounds on the external forces that can be exerted on the object. The advantage of this approach is that $J_h^T F_N$ can be calculated at the same rate as J_h—saving computation time.

A more robust F_N could be calculated by looking at the finger forces—these can be derived from the joint torques, τ, using $F_c = J_h^{-T}\tau$—and finding a null force which causes $F_c + F_N$ to lie in FC. If the grasp map has a simple form, such as the one given in the example in Section 2.2, a basis for the null space can be used to construct the set of all valid F_N. This calculation takes time but may be necessary in the case of large uncertainties.

Other grasp force calculations are discussed in [**10**] but all of these share some fundamental problems. One difficulty is that in a real-world hand the maximum motor torques that can be generated are finite. Thus, we are not guaranteed that we can apply an F_N which satisfies $F_c + F_N \in FC$ without saturating the motors. Another issue is the effect of the null force term in the presence of errors. If a large internal force term is used and, due to sensor or actuator errors, it does not actually lie in the null space of the grasp matrix, the resulting force can cause positioning errors and in the extreme case, instability.

Redundant motion. In addition to internal forces, fingers with excess degrees of freedom can have internal motions which do not cause motion of the fingers. Controllers must be extended to take into account this redundant motion. This is fundamentally no different than control of an ordinary finger except that position information is not available in redundant directions. Thus the computed torque law would become

$$F = M(q)\begin{pmatrix} \ddot{x}_d + K_v\dot{e}_x + K_p e_x \\ \ddot{y}_d + K_v\dot{e}_y \end{pmatrix} + C(q,\dot{q})\begin{pmatrix} \dot{x} \\ \dot{y} \end{pmatrix} + N(q,\dot{q}) \tag{3.29}$$

Motion specification for such a control law would be in terms of a position trajectory $x_d(\cdot)$ and a velocity trajectory $\dot{y}_d(\cdot)$.

IV. Planning

Chapter 3 was dedicated to establishing control laws under which a grasped object moved along a specified trajectory denoted $x_d(t)$. This is useful in the instance that the task involved does not necessitate a change of grasp. This is not to say that the model and control laws do not allow for fingers to roll on the surface of the object. Indeed, in this instance the motion of the finger tips described by the equations (2.31) will determine where the grasp points go and how the grasp map changes during the course of the manipulation. However, there is no explicit control of the locations of the fingertips on the surface of the object. There are however a number of applications in which an object needs to be moved while the fingers are being repositioned in some controlled fashion on the surface of the object: for instance, twirling a baton or regrasping an object for greater stability or manipulability. In this chapter we will discuss the planning of individual finger motions on the surface of an object.

4.1. Dynamic finger repositioning. In Chapter 2, we derived the kinematic equations of contact for a single finger rolling on a body. We will aggregate these into a composite equation for all of the fingers. To review the notation of Chapter 2, we recall that $g_i = x_o x_{f_i}^{-1}$ stands for the position and orientation of the ith finger ($x_{f_i} \in SE(3)$) relative to the body ($x_o \in SE(3)$). Also $\eta_i = (\alpha_{o_i}, \alpha_{f_i}, \psi_i)$ is the vector of the ith contact coordinates with $\alpha_{o_i} \in \mathbb{R}^2$ standing for the surface representation of the object, $\alpha_{f_i} \in \mathbb{R}^2$ the surface representation of the ith finger and ψ_i, the angle of contact (angle between the two orthogonal surface frames). The equations (2.31) can then be written as

$$\dot{\eta}_i = B_i(x_o, \eta_i) \begin{pmatrix} v_t \\ \omega_t \end{pmatrix} \tag{4.1}$$

where $B_i(x_o, \eta_i)$ represents the contact kinematics,

$$\begin{pmatrix} v_t \\ \omega_t \end{pmatrix} = \begin{bmatrix} M_o^{-T} \frac{\partial c_o}{\partial \alpha_o}^T & 0 \\ 0 & M_o^{-T} \frac{\partial c_o}{\partial \alpha_o}^T \end{bmatrix} \begin{bmatrix} I & \mathcal{S}(n_o) \\ 0 & \mathcal{S}(c_o) \end{bmatrix} \begin{pmatrix} v_i \\ \omega_i \end{pmatrix} \tag{4.2}$$

and

$$\begin{pmatrix} v_i \\ \omega_i \end{pmatrix} = \dot{g}_i g_i^{-1} \tag{4.3}$$

In turn, v and ω are linear functions of $v_o, \omega_o, v_{f_i}, \omega_{f_i}$ so that (4.1) may be rewritten as

$$\dot{\eta}_i = \tilde{B}_i(x_o, \eta_i) \begin{pmatrix} v_o \\ \omega_o \\ v_{f_i} \\ \omega_{f_i} \end{pmatrix} \tag{4.4}$$

Aggregating equations (4.4) for $i = 1, \ldots, k$ and using the Jacobian of the finger kinematics (2.44) we have

$$\dot{\eta} = \bar{B}(x_o, \eta) \begin{pmatrix} v_o \\ \omega_o \\ \dot{\theta} \end{pmatrix} \tag{4.5}$$

where

$$\bar{B}(x_o, \eta) = \tilde{B}(x_o, \eta) \begin{bmatrix} I & 0 \\ 0 & J_f(\theta) \end{bmatrix} \tag{4.6}$$

Also, the grasping constraint (cf. equation (2.48)) is given by

$$G^T(x_o, \theta) \begin{pmatrix} v_o \\ \omega_o \end{pmatrix} = J(x_o, \theta)\dot{\theta} \tag{4.7}$$

If the grasp map G is onto (the grasp map is stable) we can uniquely solve for (v_o, ω_o) in (4.7) as

$$\begin{pmatrix} v_o \\ \omega_o \end{pmatrix} = G^+(x_o, \theta) J(x_o, \theta)\dot{\theta} \tag{4.8}$$

Using (4.8) in (4.5) we have

$$\dot{\eta} = \bar{B}(x_o, \eta) \begin{bmatrix} G^+(x_o, \theta) J(x_o, \theta) \\ I \end{bmatrix} \dot{\theta} \tag{4.9}$$

We have determined that η is a smooth bijection of $x_o^{-1} x_f$. Further provided that the fingers do not have more than 6 degrees of freedom, x_f (locally) uniquely determines θ. Consequently (4.9) can be rewritten as

$$\dot{\eta} = \hat{B}(x_o, \eta)\dot{\theta} \tag{4.10}$$

Noting that the left hand side of (4.8) determines x_o, we now combine (4.8) and (4.10) as follows: define

$$\dot{\theta} = \dot{\theta}_1 + \dot{\theta}_2 \tag{4.11}$$

where $\dot{\theta}_1 \in \mathcal{R}(J^T(x_o, \theta))$ and $\dot{\theta}_2 \in \mathcal{N}(J(x_o, \theta))$. Further, define

$$u_1 = G^+(x_o, \theta) J(x_o, \theta) \dot{\theta}_1 x_o \tag{4.12}$$

Note that the map between u_1 and $\dot{\theta}_1$ is a bijection and let

$$\dot{\theta}_2 = K(x_o, \theta) u_2 \tag{4.13}$$

where the columns of $K(x_o, \theta)$ span the null space of $J(x_o, \theta)$. Using these two definitions it may be seen that (4.8) and (4.9) can be written as

$$\begin{aligned} \dot{x}_o &= u_1 \\ \dot{\eta} &= B_1(x_o, \eta) u_1 + B_2(x_o, \eta) u_2 \end{aligned} \tag{4.14}$$

Thus the problem of finger repositioning and body manipulation can be reformulated as the problem of steering the states (x_o, η) of the control system (4.14). In the previous chapter we neglected the u_2 and considered the problem of steering x_o (u_1 was referred to as $\dot{x}_d$). There are as many u_2 as the dimension of the null space of $J(x_o, \theta)$. Away from kinematic singularities this dimension is

$$\sum_{i=1}^{k} \max\,(n_i - m_i, 0) \tag{4.15}$$

Recall that n_i is the number of joints in the ith finger and m_i is determined by the contact type of the ith finger. The formula above represents the number of extra finger degrees of freedom available to reposition the fingers. Note also that even if $u_2 = 0$ (i.e. no extra finger degrees of freedom) it may still be possible to steer both x_o, η using the u_1 alone.

With this discussion by way of preamble, we begin a detailed of steering systems of the form

$$\dot{x} = B(x)u \tag{4.16}$$

with $x \in \mathbb{R}^n, u \in \mathbb{R}^m$. Note that our steering problem really is a steering problem on a nontrivial manifold $SE(3) \times \mathbb{R}^{5k}$ ((x_o, η) space) rather than $\mathbb{R}^n$ but we will content ourselves with a local discussion, namely on a coordinate chart of $SE(3) \times \mathbb{R}^{5k}$.

4.2. Review of Optimal Control. Following Brockett [**2**], we will review some results from optimal control. Consider control systems of the form

$$\dot{x} = B(x)u \tag{4.17}$$

Here $x \in \mathbb{R}^n$, $u \in \mathbb{R}^m$ and $B(x) \in \mathbb{R}^{n \times m}$. The optimal control problem is to minimize

$$\frac{1}{2}\int_0^1 |u|^2 dt \tag{4.18}$$

subject to the conditions $x(0) = x^i$, $x(1) = x^f$. When $m < n$, this problem is a geodesic problem with a singular Riemannian metric specified by the equations (4.17) and (4.18). From Chow's theorem, it follows that this problem has a solution for arbitrary x^i, x^f if and only if the involutive closure of the vector fields described by the columns of $B(x)$ is all of $\mathbb{R}^n$.

It is instructive to analyze some sample solutions to this problem which are in some sense canonic. We start with $n = 3$ and $m = 2$:

$$\begin{aligned} \dot{x}_1 &= u_1 \\ \dot{x}_2 &= u_2 \\ \dot{x}_3 &= x_1 u_2 - x_2 u_1 \end{aligned} \tag{4.19}$$

with $x(0) = (0, 0, 0)$ and $x(1) = (0, 0, a)$. The cost function may be written as

$$\min \frac{1}{2}\int_0^1 (\dot{x}_1^2 + \dot{x}_2^2) dt \tag{4.20}$$

with the control system (4.19) written as a constraint

$$\dot{x}_3 = x_2\dot{x}_1 - x_1\dot{x}_2 \tag{4.21}$$

From standard calculus of variations we may write the Euler-Lagrange equations with $\lambda(t)$ denoting the Lagrange multiplier as

$$\begin{aligned} \ddot{x}_1 - \lambda\dot{x}_2 &= 0 \\ \ddot{x}_2 + \lambda\dot{x}_1 &= 0 \\ \dot{\lambda} &= 0 \end{aligned} \tag{4.22}$$

Equation (4.22) establishes that $\lambda(t)$ is constant and using equation (4.19) we see that

$$\begin{bmatrix} \dot{u}_1 \\ \dot{u}_2 \end{bmatrix} = \begin{bmatrix} 0 & \lambda \\ -\lambda & 0 \end{bmatrix} \begin{bmatrix} u_1 \\ u_2 \end{bmatrix} = \Lambda \begin{bmatrix} u_1 \\ u_2 \end{bmatrix} \tag{4.23}$$

so that

$$\begin{pmatrix} u_1(t) \\ u_2(t) \end{pmatrix} = e^{\Lambda t}u(0) \tag{4.24}$$

Consequently, since $x_1(0) = x_2(0) = 0$

$$\begin{pmatrix} x_1(t) \\ x_2(t) \end{pmatrix} = \left(e^{\Lambda t} - I\right) b; \qquad b = \Lambda^{-1}u(0) \tag{4.25}$$

Further, because $x_1(1) = x_2(1) = 0$, we have that $\lambda = 2n\pi$ so as to make $e^{\Lambda} = I$. Using this fact and noticing $\dot{x}_3 = \lambda/2\,(x_1^2 + x_2^2)$ it follows that

$$x_3(1) = x_3(0) + \lambda|b|^2 = a \tag{4.26}$$

Also since the total cost is

$$\int_0^1 |u|^2 dt = \lambda^2|b|^2 \tag{4.27}$$

we see that the optimal choice of input is $\lambda = 2\pi$ and $|b|^2 = a/2\pi$ but otherwise is arbitrary. The structure of the optimal control for steering between the conjugate points $(0,0,0)$ and $(0,0,a)$ is interesting—it is sums of sines and cosines at a frequency of 2π. The frequency of the signal is dictated by the time interval. Note that the optimal input is still a combination of sines and cosines even if $x_1(1) \neq 0$ or $x_2(1) \neq 0$.

The generalization of this example to the case that $m > 2$ is as follows: consider the situation that the $\{b_i(x)\colon i = 1, \cdots, m\}$ are linearly independent for all x and also all the $m(m-1)$ first-order (first-*etage*) brackets $\{[b_i(x), b_j(x)]\colon i, j = 1, \cdots, m\}$ are linearly independent of the $b_i(x)$. The minimum dimension of the state space to allow for this possibility is $n = m + m(m-1)/2$. The canonic example of this situation is

$$\begin{aligned} \dot{x}_i &= u_i & i &= 1, \cdots, m \\ \dot{x}_{ij} &= x_i u_j - x_j u_i & i &= 1, \cdots, m \end{aligned} \tag{4.28}$$

A slightly more pleasing representation of equation (4.28) is obtained by forming the skew symmetric matrix $Y \in \mathbb{R}^{m\times m}$ with the x_{ij} as the bottom lower half (below the diagonal).

$$\begin{aligned} \dot{x} &= u && (4.29)\\ \dot{Y} &= xu^T - ux^T && (4.30) \end{aligned}$$

The Euler-Lagrange equations for (4.28) are an extension of (4.22):

$$\begin{aligned} \ddot{x} &= \Lambda\dot{x} && (4.31)\\ \dot{\Lambda} &= 0 && (4.32) \end{aligned}$$

where Λ is the skew symmetric $m \times m$ matrix of Lagrange multipliers associated with Y. Thus, as before, the optimal input u satisfies the equation

$$\dot{u} = \Lambda u \tag{4.33}$$

with $\Lambda \in \mathbb{R}^{m\times m}$ begin a constant, skew symmetric matrix. Thus $u(t)$ is a linear combination of sinusoids. The exact eigenvalues of Λ are determined by the initial and final state. In fact, if $x(0) = x(1) = 0$, $Y(0) = 0$ and $Y(1)$ a nonsingular $m \times m$ skew symmetric matrix (this requires that m is even), then it can be shown that Λ has $m/2$ sinusoids at frequencies $2\pi,\ 2\cdot 2\pi,\ \cdots,\ m/2\cdot 2\pi$. If m is odd and $Y(1)$ has only one zero eigenvalue, Λ has one zero eigenvalue and $(m-1)/2$ sinusoids at frequencies $2\pi,\ 2\cdot 2\pi,\ \cdots,\ (m-1)/2\cdot 2\pi$.

4.3. Steering of controllable systems. The results of the previous section derive the geodesics for first etage, linear systems of the form in equation (4.17). There are several ways of generalizing this further:

1. build model control systems where the vector fields

$$\begin{array}{ll} b_i & i = 1,\cdots,m \\ [b_i, b_j] & i,j = 1,\ldots,m \\ [[b_i, b_j], b_k] & i,j,k = 1,\cdots,m \end{array}$$

   are linearly independent and $n = m + m(m-1)/2 + m(m-1)(m+1)/3$.

2. understand conditions under which more general control systems can be transformed into canonical systems studied above.

3. use the results summarized thus far as inspiration to propose sinusoidal inputs at multiple frequencies which are integrally related to provide (sub-optimal) control laws to steer the systems between arbitrary initial and final conditions.

In this section we shall explore the latter possibility.

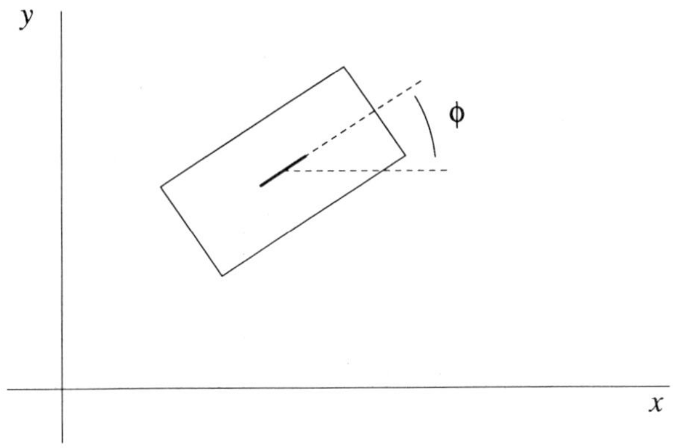

FIGURE 4.1: Steerable unicycle. The unicycle has two independent inputs: the *steering* input controls the angle of the wheel, ϕ; the *driving* input controls the velocity of the cart in the direction of the wheel. The configuration of the cart is its Cartesian location and the wheel angle.

First Etage Controllable Systems. By *first etage controllable systems* we mean systems of the form (4.17) with $m < n$, with the columns of $B(x)$ linearly independent and with

$$\operatorname{span}\{b_i(x),\ [b_i(x), b_j(x)] : i, j = 1, \ldots, m\} = \mathbb{R}^n \tag{4.34}$$

Namely, one level of Lie brackets is adequate to achieve the full tangent space for all x. We discussed the optimal control of canonical systems satisfying (4.34) in Section 4.1. Here we illustrate the application of this problem to the steering of a unicycle as shown in Figure 4.1. If u_1 denotes the driving velocity and u_2 the steering velocity the functional form of the state equations for this system is

$$\begin{aligned} \dot{x} &= \cos\phi\, u_1 \\ \dot{y} &= \sin\phi\, u_1 \\ \dot{\phi} &= u_2 \end{aligned} \tag{4.35}$$

An approximation to this system is obtained by setting $\cos\phi \simeq 1, \sin\phi \simeq \phi$ and relabelling x as x_1, y as x_3 and ϕ as x_2 to get

$$\begin{aligned} \dot{x}_1 &= u_1 \\ \dot{x}_2 &= u_2 \\ \dot{x}_3 &= x_2 u_1 \end{aligned} \tag{4.36}$$

To steer this system, we first use u_1, u_2 to steer x_1, x_2 to their desired locations; this may cause x_3 to drift. Now use $u_1 = \alpha\sin(\omega t), u_2 = \beta\cos(\omega t)$ and note that after $2\pi/\omega$ seconds, x_1 and x_2 complete a periodic trajectory and the x_3 coordinate advances by an amount equal to

$$\frac{\pi\alpha\beta}{\omega^2}$$

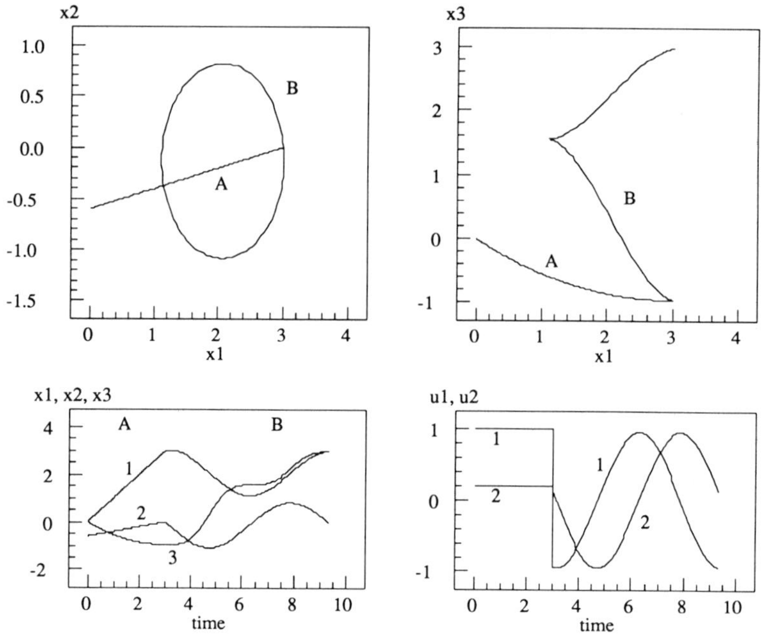

FIGURE 4.2: Sample Trajectories for the unicycle. The trajectory shown is a two stage path which moves the unicycle from (0, -.6, 0) to (3, 0, 3). The first portion of the path, labeled A, drives the x_1 and x_2 states to their desired values using a constant input. The second portion, labeled B, uses a periodic input to drive x_3 while bringing the other two states back to their desired values. The top two figures show the states versus x_1; the bottom figures show the states and inputs as functions of time.

α, β, ω can now be chosen appropriately. In order to apply this strategy to the unapproximated system (4.35) we modify the input to

$$\begin{aligned} v_1 &= \cos\phi \, u_1 \\ v_2 &= u_2 \end{aligned}$$

and relabel the states to get

$$\begin{aligned} \dot{x}_1 &= v_1 \\ \dot{x}_2 &= v_2 \\ \dot{x}_3 &= \tan x_2 \, v_1 \end{aligned} \tag{4.37}$$

As before, we steer x_1 and x_2 using v_1, v_2. To steer the third variable, we use $v_1 = \alpha \sin(\omega t), v_2 = \beta \cos(\omega t)$. Then

$$\dot{x}_3 = \tan(\frac{\beta}{\omega} \sin \omega t) \, \alpha \sin(\omega t) \tag{4.38}$$

The value of x_3 after $2\pi/\omega$ seconds is determined by the constant part of the right hand side of (4.38). The constant coefficient is given by

$$\frac{1}{2} \cdot \frac{1}{\pi} \int_{-\pi}^{\pi} \tan(\frac{\beta}{\omega} \sin\theta) \alpha \sin(\theta) d\theta$$

Sample trajectories for this scenario are shown in Figure 4.2

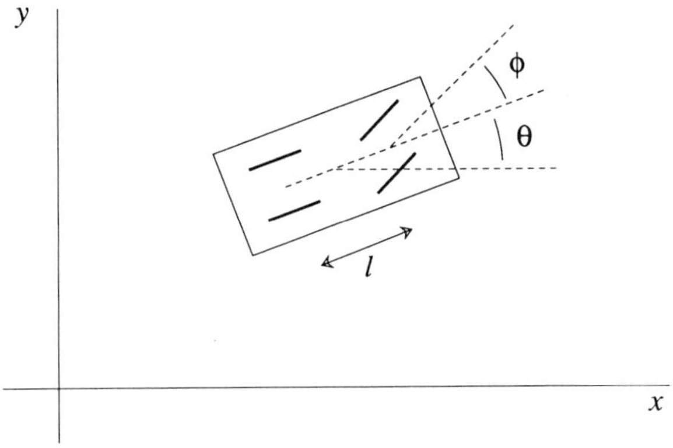

FIGURE 4.3: Front wheel drive cart. The configuration of the cart is determined by its Cartesian location, the angle the car makes with the horizontal and the steering wheel angle relative to the car body. The two inputs are the velocity of the front wheels (in the direction the wheels are pointing) and the steering velocity. The rear wheels of the cart are always aligned with the cart body and are constrained to move along the line in which they point or rotate about their center.

Second and Higher Etage Controllable Systems. Systems (4.17) are said to be *second etage controllable* if

$$\text{span}\{b_i(x),\ [b_i(x), b_j(x)],\ [b_i, [b_j, b_k]] : i, j, k = 1, \ldots, m\} = \mathbb{R}^n$$

An example of such a system is a front wheel drive cart of the form shown in Figure 4.3. As in the case of the previous example u_1 is the driving velocity and u_2 the steering velocity. The equations of this cart are

$$\begin{aligned} \dot{x} &= \cos\theta \cos\phi \ u_1 \\ \dot{y} &= \sin\theta \cos\phi \ u_1 \\ \dot{\phi} &= u_2 \\ \dot{\theta} &= \tfrac{1}{l} \sin\phi \ u_1 \end{aligned} \tag{4.39}$$

The form of the equations shows that when $\phi = \pi/2$, the cart cannot be driven forward. As in the previous section an approximation to this system is instructive. Relabelling the variables x, y, ϕ, θ as x_1, x_4, x_2, x_3, setting $l = 1$ and approximating sines and cosines as before yields

$$\begin{aligned} \dot{x}_1 &= u_1 \\ \dot{x}_2 &= u_2 \\ \dot{x}_3 &= x_2 u_1 \\ \dot{x}_4 &= x_3 u_1 \end{aligned} \tag{4.40}$$

Note that $\text{span}\{b_1, b_2, [b_1, b_2], [b_1, [b_1, b_2]]\} = \mathbb{R}^n$ so that system is a second etage system. It is also easy to verify that this condition also holds for the original system. Steering the states x_1, x_2, x_3 of (4.40) is immediate from the previous section. To steer x_4 note that if

$$u_1 = \alpha \cos \omega t \ , \ \ u_2 = \beta \cos 2\omega t$$

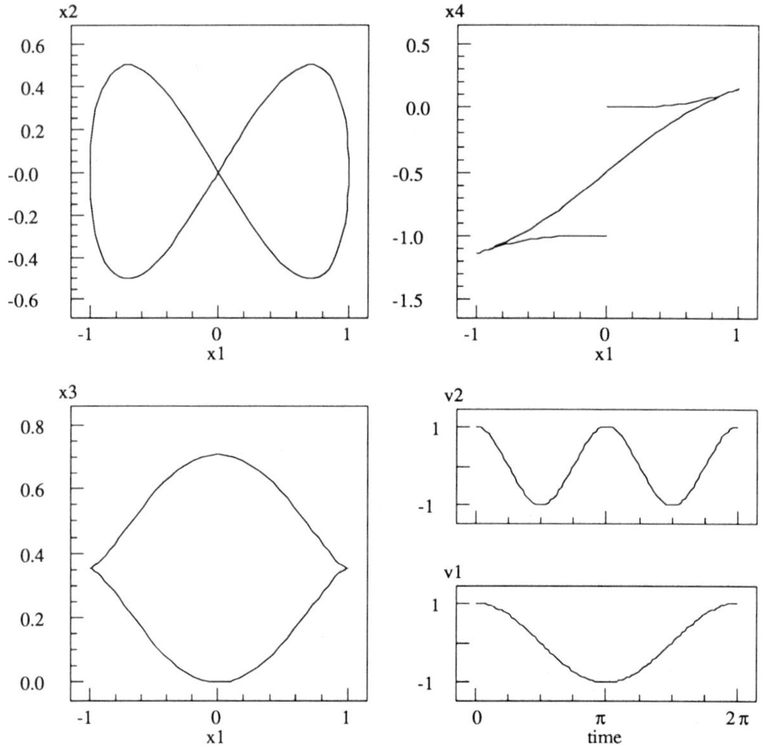

FIGURE 4.4: Trajectories for a simple cart. The trajectory shown illustrates motion in the x_4 direction. Periodic inputs are used to generate periodic trajectories in the first three states while giving an open trajectory in the last state.

then x_1, x_2 and x_3 are all periodic and return to their initial values after $2\pi/\omega$ seconds. Also

$$x_3 = -\frac{\alpha\beta}{4\omega^2}\cos\omega t \quad - \quad \frac{\alpha\beta}{12\omega^2}\cos 3\omega t$$

so that it too is periodic. Finally the increment in x_4 is given by

$$-\frac{\pi\alpha^2\beta}{4\omega^3}$$

To carry this development through for the unapproximated system define $v_1 = u_1\cos\theta\cos\phi$ and $v_2 = u_2$. Then with the same relabelling as before, the equations become

$$\begin{aligned} \dot{x}_1 &= v_1 \\ \dot{x}_2 &= v_2 \\ \dot{x}_3 &= \frac{\tan(x_2)}{\cos(x_3)} v_1 \\ \dot{x}_4 &= \tan x_3 v_1 \end{aligned} \tag{4.41}$$

We refer to such systems as *triangular* but not *strictly triangular* since $\dot{x}_3$ depends on x_3. By approximating $\cos x_3$ by 1, the equations become strictly triangular;

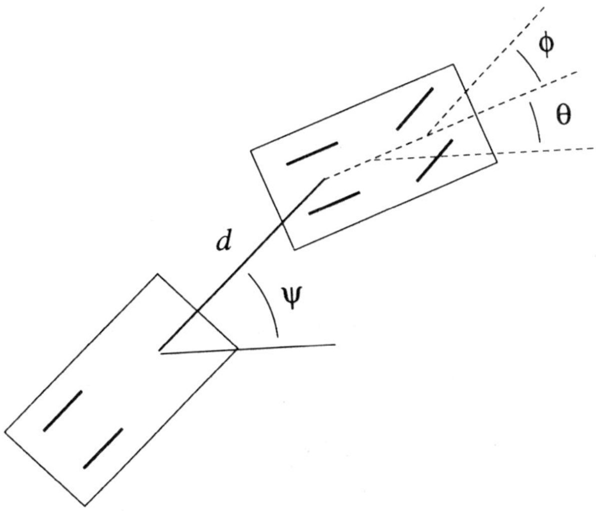

FIGURE 4.5: Front wheel drive cart with trailer. The trailer configuration is described the the angle the trailer makes with the horizontal, ψ. The rear wheels of the trailer are fixed and constrained to move along the line in which they point or rotate about their center. The inputs to the system are the inputs to the cart: the driving velocity (of the front wheels) and the steering velocity. This system is an example of a third etage system; higher etage systems can be generated by adding extra trailers.

using $v_1 = \alpha \cos \omega t$, $v_2 = \beta \cos 2\omega t$ we can solve for the Fourier series coefficients of x_1, x_2, x_3 and x_4. Note that only the Fourier coefficient corresponding to the zero frequency is needed to get the change in x_4 after one time period.

To summarize, it is easy to see that for higher etage (than 2) controllable systems one can use simple Fourier series techniques to steer the systems using as inputs integrally related sinusoids provided that they are strictly triangular in the sense discussed above. To steer the variable corresponding to the kth etage it is possible to use frequencies ω and $k\omega$ in the two inputs. The Lissajous figures that are obtained from the phase portraits of the different variables are quite instructive. Consider the Figure 4.4, which is the system of (4.41) with $\cos x_3$ replaced by 1 and the inputs $v_1 = \alpha \cos \omega t$, $v_2 = \beta \cos 2\omega t$. The upper left plot is the Lissajous figure for x_1, x_2 (two loops); The lower left plot is the corresponding figure for x_3, x_1 (one loop) and the open curve in x_4,x_1 shows the increment in the x_4 variable. The very powerful implication here is that the *Lie bracket directions correspond to rectification of harmonic periodic motions of the driving vector fields and the harmonic relations are determined by the etage of controllability desired.* This point has also been made rather elegantly by Brockett **[3]** in the context of the rectification of mechanical motion.

Open Problems and Nontriangular Higher Etage Systems. Consider the kinematic equations for a front wheel drive cart with a trailer as shown in Figure 4.5. The kinematic equations are those of the car with an additional equation to describe the angle of the trailer:

$$\begin{aligned} \dot{x} &= \cos\theta \cos\phi \; u_1 \\ \dot{y} &= \sin\theta \cos\phi \; u_1 \\ \dot{\theta} &= \tfrac{1}{l} \sin\phi \; u_1 \\ \dot{\phi} &= u_2 \\ \dot{\psi} &= \tfrac{1}{d} \sin(\theta - \psi) \cos\phi \; u_1 \end{aligned} \tag{4.42}$$

It may be verified that

$$\operatorname{span}\{b_1,\, b_2,\, [b_1, b_2],\, [b_1, [b_1, b_2]]\,,\, [b_1, [b_1, [b_1, b_2]]]\} = \mathbb{R}^5$$

It is also not difficult to see that with k trailers we need Lie brackets up to the $k+2$ etage to guarantee controllability. Also, it may be seen that after redefining the inputs the system is only triangular rather than strictly triangular so that the harmonic analysis techniques of the previous section cannot be applied even though numerical simulation suggests that sinusoids of integer multiples are useful to steer along the direction of the jth Lie bracket. The full theory for these systems is as yet incomplete.

4.4. Dynamic finger repositioning revisited. We now reconsider the case of a multifingered hand grasping an object. The equations of motion are given by equation (4.14) which we reproduce here:

$$\begin{aligned} \dot{x}_o &= u_1 \\ \dot{\eta} &= B_1(x_o, \eta) u_1 + B_2(x_o, \eta) u_2 \end{aligned} \tag{4.43}$$

Recall that these equations were obtained by attaching a controller to the system and letting u_1 reflect the desired object velocity and u_2 parameterize the internal motion of the system.

The general case of finding $u_1(t)$ and $u_2(t)$ such that the object and the fingers move from an initial to final position (while maintaining contact) can be very difficult. $B_1(x_o, \eta)$ and $B_2(x_o, \eta)$ are rarely in any of the simple forms that we have considered thus far. We point out two interesting special cases:

1. If the hand has no redundant degrees of freedom (i.e., B_2 is not present) then it might be possible to move to an arbitrary location/grasp using only u_1. Moving just the contact location requires a carefully chosen closed loop path in x_o.

2. If we have redundant degrees of freedom, then we can move the fingers along the object while keeping the object position fixed ($\dot{x}_o = u_1 = 0$). In this case we use only the vector fields in B_2 to move the fingers.

In the second case, it is sufficient to study the control of a single finger since the fingers are decoupled if the object is held fixed. This situation has been studied by Li and Canny [**9**] and we review some examples from that paper in the context of steering controllable systems.

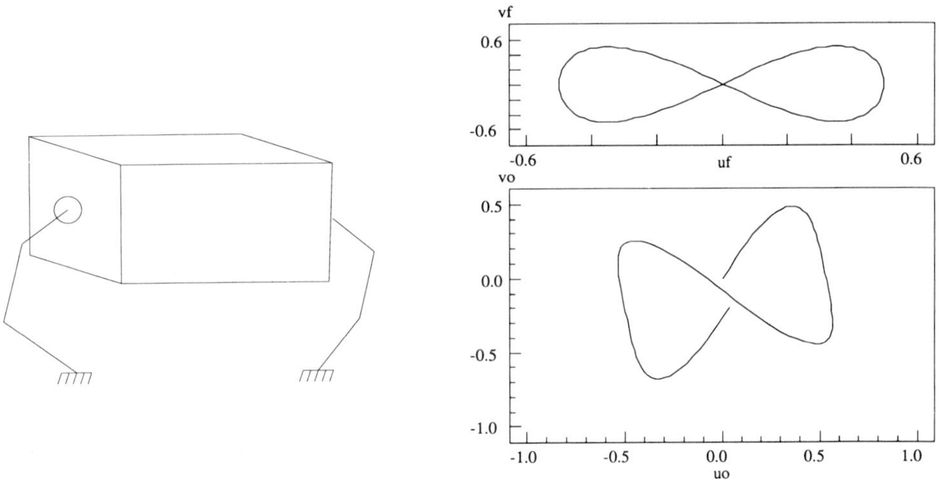

FIGURE 4.6: Steering applied to a multifingered hand. We consider the motion of a finger with a spherical tip on a polyhedral object (left). The plots to the right show a trajectories which move a finger down the side of an object. The location of the contact on the finger is unchanged (upper graph), while the location of the contact on the face of the object undergoes a net y displacement (lower graph).

Examples. Consider the case of a single spherical finger rolling on a plane. The control kinematics were derived in Section 2.3:

$$\dot{\eta} = \begin{pmatrix} \dot{u}_f \\ \dot{v}_f \\ \dot{u}_o \\ \dot{v}_o \\ \dot{\psi} \end{pmatrix} = \begin{pmatrix} 1 \\ 0 \\ \rho\cos\psi \\ -\rho\sin\psi \\ 0 \end{pmatrix} \omega_1 + \begin{pmatrix} 0 \\ \sec u_f \\ -\rho\sin\psi \\ -\rho\cos\psi \\ -\tan u_f \end{pmatrix} \omega_2 \tag{4.44}$$

For simplicity, we assume that we control ω_1 and ω_2 directly. It can be verified that the system is second etage controllable and that by a change of input variables we can put this system into strictly triangular form. Furthermore, the approximate version of this system is given by

$$\begin{aligned} u_f: \quad \dot{x}_1 &= u_1 \\ v_f: \quad \dot{x}_2 &= u_2 \\ -\psi: \quad \dot{x}_3 &= x_1 u_2 \\ u_o - \rho u_f: \quad \dot{x}_4 &= x_3 u_2 \\ v_o + \rho v_f: \quad \dot{x}_5 &= x_3 u_1 \end{aligned} \tag{4.45}$$

This is identical to the approximat cart kinematics in equation (4.40) with the addition of an extra state. Using the same techniques as before, we can construct paths using integrally related sinusoids and apply these sinusoids to the full nonlinear system in equation (4.44). An example of a path which moves a finger vertically down the side of a planar object is shown in Figure 4.6.

A more challenging example considered by Li and Canny is that of moving a spherical finger on a spherical object. It may be verified that the system is controllable except when the object and finger radii are identical (in this case the rolling constraint becomes holonomic). The contact kinematics, from equation (2.43) are

$$\dot{\eta} = \begin{pmatrix} \frac{1}{1+\rho} \\ 0 \\ \frac{\rho}{1+\rho}\cos\psi \\ -\frac{\rho}{1+\rho}\sin\psi\sec u_o \\ \frac{\rho}{1+\rho}\sin\psi\tan u_o \end{pmatrix}\omega_1 + \begin{pmatrix} 0 \\ \frac{1}{1+\rho}\sec u_f \\ -\frac{\rho}{1+\rho}\sin\psi \\ -\frac{\rho}{1+\rho}\cos\psi\sec u_o \\ \frac{\rho}{1+\rho}(\cos\psi\tan u_o - 1/\rho\ \tan u_f) \end{pmatrix}\omega_2 \quad (4.46)$$

We see that this system is not strictly triangular (ψ depends on u_o) and hence requires a more sophisticated approach. Motion for this particular system can be constructed using the techniques described by Li and Canny due to the special choice of object and finger shapes. Motion planning for more general choices of finger and object shapes is still unsolved.

Acknowledgements. We would like to thank Roger Brockett for introducing us to his work on singular Riemannian metrics (Chapter 4), Li Zexiang for making the connections between regrasping and steering, and John Canny, Jean-Paul Laumond and the denizens of the robotics lab for several useful discussions.

Bibliography

1. A. K. Bejczy. Robot arm dynamics and control. Technical Report 33-699, Jet Propulsion Laboratory, 1974.

2. R. W. Brockett. Control theory and singular riemannian geometry. In *New Directions in Applied Mathematics*, pages 11–27. Springer-Verlag, New York, 1981.

3. R. W. Brockett. On the rectification of vibratory motion. *Sensors and Actuators*, 1989. (to appear).

4. A. Cole, J. Hauser, and S. Sastry. Kinematics and control of multifingered hands with rolling contact. In *IEEE International Conference on Robotics and Automation*, pages 228–233, 1988.

5. J. Kerr. *An Analysis of Multi-fingered Hands.* PhD thesis, Stanford University, Department of Mechanical Engineering, 1984.

6. O. Khatib. A unified approach for motion and force control of robot manipulators: The operational space formulation. *IEEE Journal on Robotics and Automation*, RA-3(1):43–53, 1987.

7. O. Khatib. Augmented object and reduced effective inertia in robot systems. In *Automatic Control Conference*, 1988.

8. D. Koditschek. Natural motion for robot arms. In *IEEE Control and Decision Conference*, pages 733–735, 1984.

9. Z. Li and J. Canny. Motion of two rigid bodies with rolling constraint. *IEEE Transactions on Robotics and Automation*, 6(1):62–71, 1990.

10. Z. Li, P. Hsu, and S. Sastry. On kinematics and control of multifingered hands. In *IEEE International Conference on Robotics and Automation*, pages 384–389, 1988.

11. J. Y. S. Luh, M. W. Walker, and R. P. Paul. Resolved acceleration control of mechanical manipulators. *IEEE Transactions on Automatic Control*, AC-25, 1980.

12. D. J. Montana. The kinematics of contact and grasp. *International Journal of Robotics Research*, 7(3):17–32, 1988.

13. R. Murray and S. Sastry. Control experiments in planar manipulation and grasping. In *IEEE International Conference on Robotics and Automation*, pages 624–631, 1989.

14. V-D. Nguyen. Constructing force-closure grasps. *International Journal of Robotics Research*, 7(3):3–16, 1988.

15. R. M. Rosenberg. *Analytical Dynamics of Discrete Systems.* Plenum Press, New York, 1977.

16. N. Sadegh. *Adaptive Control of Mechanical Manipulators: Stability and Robustness Analysis.* PhD thesis, Department of Mechanical Engineering, University of California, Berkeley, California, 1987.

17. J. E. Slotine and W. Li. On the adaptive control of robot manipulators. *International Journal of Robotics Research*, 6:49–59, 1987.

DEPARTMENT OF ELECTRICAL ENGINEERING AND COMPUTER SCIENCE,
UNIVERSITY OF CALIFORNIA, BERKELEY CA 94720

Proceedings of Symposia in Applied Mathematics
Volume **41**, 1990

Planning and Executing Robot Assembly Strategies in the Presence of Uncertainty

Bruce R. Donald[1]

Abstract

Robot control systems are subject to significant uncertainty and error. Typical robots are also equipped with sensors—force sensors, kinesthetic position sensors, tactile sensors, vision, and so forth. However, these sensors are also subject to significant uncertainty. Finally, the geometrical models of the robot and the environment (parts, obstacles, etc.) cannot be exact—they are accurate only to manufacturing tolerances, or to the accuracy of the sensors used to acquire the models. Uncertainty is an absolutely fundamental problem in robotics, and plans produced under the assumption of no uncertainty are meaningless. What is needed is a principled theory of planning in the presence of uncertainty. Such a theory must not only be computational, but must also take uncertainty into account *a priori*. In motion planning with uncertainty, we exploit compliant motion—sliding on surfaces—in order to effect a "structural" reduction in uncertainty. Such compliant motion plans can be synthesized from a computational analysis of the geometry of the holonomic constraints.

We will present a precise framework for motion planning with uncertainty. In particular, given geometric bounds on the uncertainty in sensing and control, we develop algorithms for generating and verifying compliant motion strategies that are guaranteed to succeed as long as the sensing and control uncertainties lie within the specified bounds. The first results in this theory begin with Lozano-Pérez, Mason, and Taylor [LMT], with subsequent contributions by Mason [Ma2], Erdmann [E], Donald [D], and others. This research has led to a theoretical computational framework for motion planning with uncertainty, which we explore in this focused survey paper.

1980 *Mathematics Subject Classification* (1985 *Revision*). Primary 68T05.

[1]This paper describes research done in the Computer Science Robotics Laboratory at Cornell University. Support for our robotics research is provided in part by the National Science Foundation under grant No. IRI-8802390 and by a Presidential Young Investigator award, and in part by the Mathematical Sciences Institute.

1 A Geometric Approach to Robot Planning

1.1 Introduction

For the past seven years, I have been interested in building robot planning systems that can function at the task-level. A task-level specification of a robot plan might have the form, *Put together this disk rotor assembly.* The planner is given geometric models of the parts, and geometric or analytic models of the robot dynamics. Beyond this, the specification, or input to the planner does not mention the specific kinematic and dynamic constraints that the robot must obey; these are determined by the planner using geometrical computation. Typically, there are additional constraints that the planner must also obey, for example, *Construct a robot plan that is robust in the face of uncertainty and error. Use compliant motion where appropriate to reduce uncertainty.* The goal of a task-level planner is to take a task-level specification and to produce a runnable robot program—one which is fully specified in terms of force-control, kinematics, and dynamics—that can accomplish the task.

Major advances in task-level planning can enable robotics to achieve its full potential in the assembly domain. Today, even existing robots cannot be exploited to their full capacity. For example, assembly tasks require compliant motion; however, compliance requires force-control, and such force-control motion strategies are quite difficult for humans to specify. Furthermore, robot assembly programs are very sensitive to the details of geometry. Finally, reprogramming a general purpose robot for a new assembly task can take time on the order of man-months. For these reasons, we have been working on the automatic synthesis of motion strategies for robots. Much work is required to reduce such current task-level technologies to practice.

Research in task-level planning is often characterized as *theoretical robotics.* There are several reasons for this; the first is that much of the work has been concerned with constructing a *theory* of planning. In other words, the computational problem "task-level planning" is not well-specified. Much of our work lies in specifying the computation precisely. Second, given some sort of decomposition of task-level planning into "planning problems", one is immediately driven to ask, *What are the algorithms for these problems? Can plans, in general, be computed? How efficiently can planning algorithms run?* Historically, the nature of these questions has led researchers to apply tools from theoretical computer science, computational geometry, and algebra.

Recently, a great deal of attention has been focused on a particular robotics problem, called the find-path, or generalized movers' problem. In this problem, we ask the purely kinematic question, can a robot system be moved from one configuration to another, without colliding with obstacles? (See figs. 1, 2). This is a nicely-defined mathematical problem, and, after much research, at this point its computational complexity is precisely known.

In fact, the neatness of this problem is deceptive, so much so that this formal problem

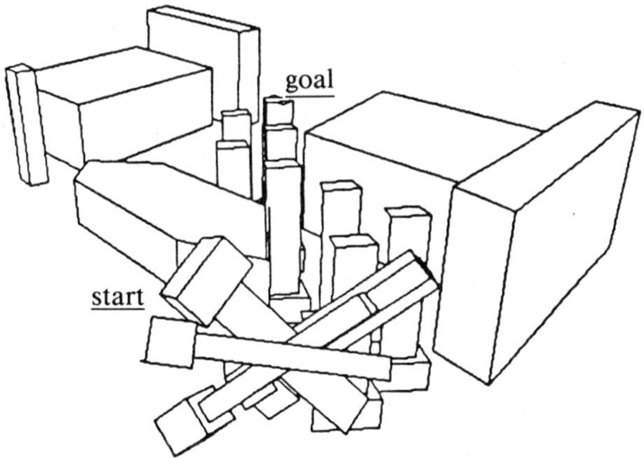

Figure 1: An example of a solution path for the classical find-path, or Movers' problem. This illustration is a "time-lapse" picture of a path found for a hammer-shaped object through a maze of polyhedral obstacles. From [Donald, 87a].

has even been called "the" motion planning problem. From a task-level viewpoint, there is much hidden in the statement "Can the robot system be moved...?" Specifically, the find-path problem assumes that the robot has a perfect control system that can exactly execute the plan, and that the geometric and analytic models of the robot and obstacles are exact.

In reality, of course, robot control systems are subject to significant uncertainty and error. Typical robots are also equipped with sensors—force sensors, kinesthetic position sensors, tactile sensors, vision, and so forth. However, these sensors are also subject to significant uncertainty. Finally, the geometrical models of the robot and the environment (parts, obstacles, etc.) cannot be exact—they are accurate only to manufacturing tolerances, or to the accuracy of the sensors used to acquire the models. Uncertainty is not a mere engineering detail; in particular, it is characteristically impossible to "patch" these perfect plans in such a way that they will function once uncertainty comes into play. Uncertainty is an absolutely fundamental problem in robotics, and plans produced under the assumption of no uncertainty are meaningless. What is needed is a principled theory of planning in the presence of uncertainty. Such a theory must not only be computational, but must also take uncertainty into account *a priori*. The overlap with exact motion planning algorithms can be stated roughly as follows: exact kinematic planning algorithms provide a computational-geometric theory of holonomic constraints. In motion planning with uncertainty, we exploit compliant motion—sliding on surfaces—in order to effect a

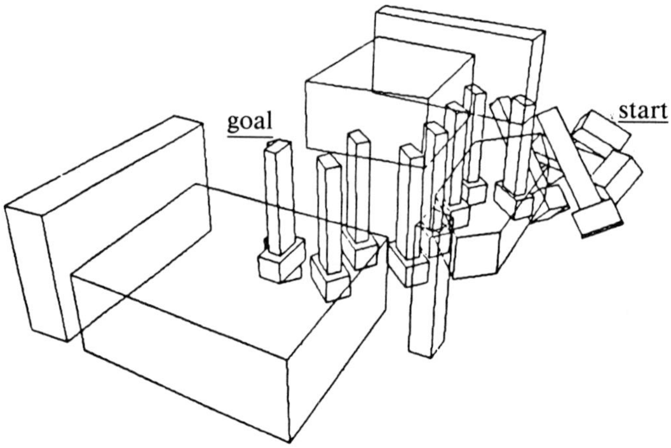

Figure 2: A different view of the solution path for the Movers' problem in fig. 1, with the obstacles "transparent" to allow us to view the rotations better.

"structural" reduction in uncertainty. Such compliant motion plans can be synthesized from a computational analysis of the geometry of the holonomic constraints.

1.2 Uncertainty and Compliant Motion

Robots are subject to the following kinds of uncertainty:

1. Inaccuracy and errors in sensing,

2. Inaccuracy and errors in control,

3. Uncertainty about the geometry of the environment.

The last (3) is called "model error", and has received little previous attention. Model error arises because, in general, a robot can have only approximate knowledge of the shape and position of objects in the environment.

We now ask the question:

- How can robots plan and execute tasks (for example, a mechanical assembly using compliant motion) in the presence of these three kinds of uncertainty?

This is perhaps the most fundamental problem in robotics today. We call it the problem of *motion planning with uncertainty.*

In motion planning with uncertainty, the objective is to find a plan which is guaranteed to succeed even when the robot cannot execute it perfectly due to control and sensing uncertainty. With control uncertainty, it is impossible to perform assembly tasks which involve sliding motions using position control alone. To successfully perform assembly tasks, uncertainty must be taken into account, and other types of control must be employed which allow *compliant motion.*

Compliant motion occurs when a robot is commanded to move into an obstacle, but rather than stubbornly obeying its motion command, it complies to the surface of the obstacle. Work on compliant motion[2] attempts to utilize the task geometry to plan motions that reduce the uncertainty in position by maintaining sliding contact with a surface. Plans consisting of such motions can be designed to exploit the geometry of surfaces around the goal to guide the robot. By computing "preimages"[3] of a geometrical goal in configuration space, guaranteed strategies can be synthesized geometrically: We call this a *geometrical theory of planning.* The first results in this theory begin with Lozano-Pérez, Mason, and Taylor (or [LMT]), with subsequent contributions by Mason [Ma2], Erdmann [Erdmann] and Donald [D]. This research has led to a theoretical computational framework for motion planning with uncertainty, which we denote [LMT,E,D]. See [Buc, EM, Bro, CR, Can89a, FHS, LLS] for other allied work.

The [LMT,E,D] framework begins by observing that the use of active compliance enables robots to carry out tasks in the presence of significant sensing and control errors. Compliant motion meets external constraints by specifying how the robot's motion should be modified in response to the forces generated when the constraints are violated. For example, contact with a surface can be guaranteed by maintaining a small force normal to the surface. The remaining degrees of freedom (DOF)—the orthogonal complement of the normal-space—can then be position-controlled. Using this technique, the robot can achieve and retain contact with a surface that may vary significantly in shape and orientation from the programmer's expectations. Generalizations of this principle can be used to accomplish a wide variety of tasks involving constrained motion, e.g., inserting a peg in a hole, or following a weld seam. The specification of particular compliant motions to achieve a task requires knowledge of the geometric constraints imposed by the task. Given a description of the constraints, choices can be made for the compliant motion parameters, e.g., the motion freedoms to be force controlled and those to be position controlled. It is common, however, for position uncertainty to be large enough so that the programmer cannot unambiguously determine which geometric constraints hold at any instant in time. For example, the possible initial configurations for a peg in hole strategy (see fig. 3) may be "topologically" very different, in that different surfaces of the peg and hole are in contact. Under these circumstances, the programmer must employ a combined strategy of force and position control that guarantees reaching the desired final configuration from all the likely initial

[2]See [Ma] for an introduction and survey.

[3]The *preimage* of a goal [LMT] is the set of configurations from which a particular commanded compliant motion is guaranteed to succeed.

configurations. We call such a strategy a *motion strategy.*

Motion strategies are quite difficult for humans to specify. Furthermore, robot programs are very sensitive to the details of geometry. For this reason, we have been working on the automatic synthesis of motion strategies for robots.

Note that compliant motion planning with uncertainty is significantly different from motion planning with perfect sensing and control along completely-known configuration space obstacle boundaries [Kou, HW, BK]. The first difference is physical:

- From a practical point of view, the motion-in-contact plans generated under the assumption of perfect control cannot ever be executed by a physical robot using position control alone.

The second difference is combinatorial:

- The planning of motions in contact with perfect control has the same time-complexity as planning free-space motions; that is, it can be done in time $O(n^r \log n)$ for r degrees of freedom and n faces or surfaces in the environment [C1]; the exponent is worst-case optimal. However, prior to [Donald 87b,88a], there are no upper bounds for planning compliant motions *with* uncertainty. However, for r fixed at 3, the problem is hard for non-deterministic exponential time [CR]. We showed in [Donald 87b,88a] that the planar case, somewhat surprisingly, turns out to be tractable.

The physical difficulty in fact motivates our work. This experimental viewpoint is manifest in [Donald 87b,88b; Jennings, Donald and Campbell]. It is also possible to adopt a more theoretical perspective; in [Donald 88a; Briggs 89] we concentrate on the geometric and combinatorial aspects of the problem.

1.3 Example: Synthesizing Guaranteed Plans

Error Detection and Recovery (EDR) strategies arose as a response to certain inadequacies in the "guaranteed success" planning model. Hence, in order to study EDR, it is first necessary to understand the structure of guaranteed compliant motion strategies, and how they may be synthesized. To this end, in these notes, we wish to give a flavor for the kinds of issues that arise in planning guaranteed motion strategies under uncertainty. We examine a very special case—the planar polygonal case. (That is, the workspace environment ("parts" and "obstacles") are polygonal). First, we will briefly develop a simple dynamic model that is adequate for this situation. Next, we carefully define the computational problem of synthesizing planar guaranteed strategies under uncertainty in sensing and control, and generalize our definition to include uncertainty in the shape of the parts and obstacles. Finally, we hint at the types of computational techniques employed.

For this very simple planning problem, we find that motion plans with a simple structure suffice. This permits us to illustrate by a specific example the situation that we investigated in [Donald 87b, 88a, 88b, 89] for fairly arbitrary plans. Together with the problem of constructing guaranteed compliant motion strategies under uncertainty, we will consider a rather restricted class of strategies, namely those that terminate by sticking on a surface.

1.4 Dynamic Model

Compliant motion is only possible with certain dynamic models. We will employ the generalized damper model [W, Ma]. We assume that the environment is polyhedral, and that it describes the configuration space of the robot, so that the robot is always a point. The planned path consists of r successive motions in directions $v_1, \ldots, v_r$. Each motion terminates when it sticks, due to coulomb friction, on some surface in the environment. Because of control uncertainty, however, the robot cannot move with precisely velocity v_i on the ith motion. Instead, it moves with velocity v_i^{free}, which lies in a cone of velocities $B_{ec}(v_i)$ about v_i. The boundaries of the cone form an angle of ϵ_c with v_i. ϵ_c is called the *control uncertainty*, and $B_{ec}(v_i)$ the *control uncertainty cone about* v_i. It is, in fact, equivalent to regard ϵ_c as specifying that v_i^{free} lies within a ball about v_i in velocity space.

For a compliant motion, the robot moves along an obstacle surface with a sliding velocity v_i^{slide} which is the projection onto the surface of the obstacle of some v_i^{free} in $B_{ec}(v_i)$. Under generalized damper dynamics, the motion of a polyhedral robot without rotations is completely specified by the motion of its reference point in configuration space. See fig 3.

The ith motion terminates by sticking on a surface when the velocity v_i^{free} in $B_{ec}(v_i)$ points into the negative coulomb friction cone on a surface (see fig. 4). Thus sticking on a surface can be non-deterministic. We will assume that motion i can terminate on any reachable surface for which some velocity $v_i^{free} \in B_{ec}(v_i)$ is inside the negative friction cone. Sticking termination is motivated by the fact that a robot with a force-sensing wrist can easily recognize sticking and robustly terminate the motion.

To test whether sticking is possible on some set of (say, goal) edges, we simply perform a geometric cone intersection on each edge. Sticking is possible when the intersection of the cone of velocity uncertainty and the negative friction cone have a non-trivial intersection. Since determining the possibility (or necessity) of sticking reduces to a simple cone intersection, which may be done in constant time per edge, in this preface we will focus on the more difficult issue of computing reachability. Representing friction in our planar polygonal configuration space is easy; see fig. 4. However, the more general question of representing friction in configuration spaces with rotations is subtle; see [Erdmann, BRS].

While robust implementation of generalized damper dynamics is still a research issue, in our robotics laboratory we have recently implemented an experimental force-control system with this dynamic model to test our geometrical planning theories.

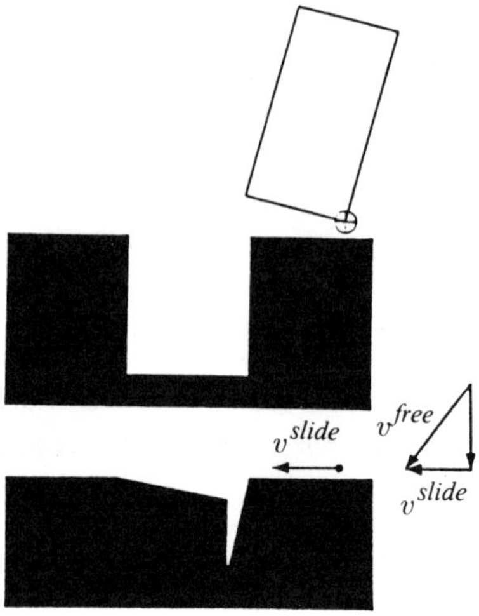

Figure 3: (a) Peg in hole environment. (b) Configuration space, showing the motion of the reference point during compliant motion.

1.5 Definitions

We will regard the goal region G as a polyhedral region in configuration space. Since in general we cannot precisely know the initial configuration of the robot, we will also assume that the start region R is some polyhedral region in configuration space.

We now pose three problems (see fig. 6):

Problem 1: One-Step Compliant Motion Planning with Uncertainty. Given a polyhedral start region R, a polyhedral environment $\mathcal{P}$ of n vertices, control uncertainty ϵ_c, coefficient of friction μ, and a polyhedral goal G, find one commanded motion direction v such that under v, all possible motions from R terminate by sticking in G.

Problem 2: One-Step Compliant Motion Verification. Given $(R, \mathcal{P}, \epsilon_c, \mu, G)$ and v, verify that under v, all possible motions from R terminate by sticking in G.

Problem 3: Compliant Motion Planning with Uncertainty Given $(R, \mathcal{P}, \epsilon_c, \mu, G)$, and an integer r, find a sequence of r motions such that each motion terminates in sticking, and the final motion terminates in the goal. Or, if no such r-step strategy exists, then say so.

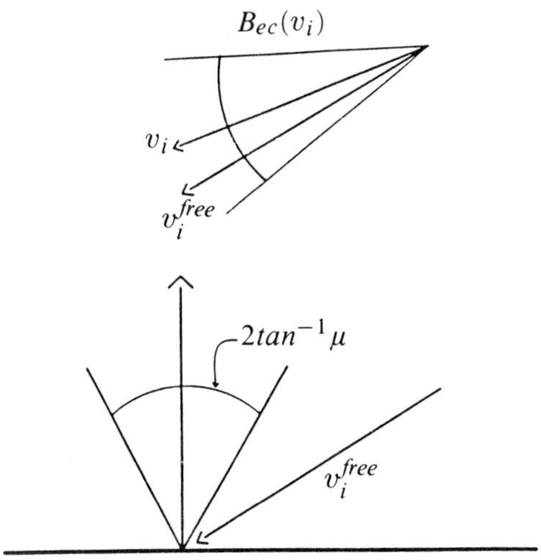

Figure 4: The control uncertainty cone B_{ec} about a nominal commanded velocity v_i. The coulomb friction cone edge makes an angle of $\tan^{-1}\mu$ with the normal. The velocity v_i^{free} will result in sliding on the surface.

1.6 Model Error

We now introduce model error into the picture. How can compliant motion strategies be synthesized in the presence of sensing, control, and geometric model error, such that the strategies are guaranteed to succeed so long as the errors lie within the specified bounds? As an example, consider a peg-in-hole assembly with sensing and control uncertainty, with toleranced parts. We wish to synthesize a compliant motion strategy that is guaranteed to succeed so long as the parts lie within the specified tolerances, and the sensing and control errors lie within the specified bounds.

We can now state our fourth problem. We parameterize our family of geometries by the parameter space $\alpha_1, \ldots, \alpha_k$, so that $\mathcal{P}(\alpha_1, \ldots, \alpha_k)$ denotes a particular geometry, and each α_i lies within some given interval. Note that this parameterization need not be "continuous". We wish to find a plan which will succeed for all geometries $\mathcal{P}(\alpha_1, \ldots, \alpha_k)$.

Problem 4: Compliant Motion Planning with Uncertainty and Geometric Model Error Given $(R, \mathcal{P}(\alpha_1, \ldots, \alpha_k), \epsilon_c, \mu, G)$, and an integer r, find a sequence of r motions such that, for all possible values of $\alpha_1, \ldots, \alpha_k$, each motion terminates in sticking, and the final motion terminates in the goal. Or, if no such r-step strategy exists, then say so.

We have begun an attack on this problem by introducing additional dimensions to the configuration space; each dimension represents a way in which the parts could parametrically vary. We termed the product space of the motion degrees of freedom and the geometric model variational dimensions "generalized configuration space" and showed how to compute "preimages" [LMT,E] of a geometrical goal in this generalized configuration space. The preimage of a goal is the set of (generalized) configurations from which a particular commanded compliant motion is guaranteed to succeed. Using this technique, we have developed a framework for computing motion strategies that are guaranteed to succeed in the presence of sensing, control, and geometric model uncertainty. The motion strategies comprise sensor-based gross motions, compliant motions, and simple pushing motions.

1.7 Backprojections

We now sketch, for a specific example, a computational approach to the compliant motion planning problem with uncertainty. In [Donald 87b, 88a;b, 89] we have mounted a more systematic attack using similar, albeit considerably generalized methods.

Erdmann [Erdmann] has shown that in the plane, when G is a single edge of the polygonal environment $\mathcal{P}$, then the one-step verification problem (2) can be done in time $O((n+c)\log n)$, where c is the number of intersections encountered by a planar arrangement algorithm.

Erdmann's algorithm makes use of *backprojections*, which he defined as a simplified case of the [LMT] notion of geometrical preimages. The question of goal reachability from a start region can be reduced to deciding the containment of the start region within the backprojection of the goal.

The backprojection $B_\theta(G)$ of a goal G (with respect to a commanded velocity v_θ^*) consists of those configurations guaranteed to enter the goal (under v_θ^*).[4] That is, the backprojection is the set of all positions from which all possible trajectories consistent with the control uncertainty are guaranteed to reach G. See fig. 5. The terms "preimage" and "backprojection" come from viewing motions as "mappings" between subsets of configuration space. Hence the backprojection of a goal is the set of configurations from which a particular commanded compliant motion is guaranteed to succeed. [LMT] envisioned a back-chaining planner that recursively computes preimages of a goal region. Successive subgoals are attained by motion strategies. Each motion terminates when all sensor interpretations indicate that the robot must be within the subgoal.

See fig. 6. Here is the key point about backprojections: Given $(R, \mathcal{P}, \epsilon_c, \mu, G, v_\theta^*)$, the one-step verification problem (2) reduces to testing set containment, i.e., that

$$R \subset B_\theta(G).$$

Erdmann showed that when G is a single edge of a planar environment $\mathcal{P}$, then $B_\theta(G)$ has size $O(n)$ and can be computed as follows (see fig. 7):

[4]The star * denotes the ideal, or perfect control velocity. Henceforth, we will typically identify a commanded motion v_θ^* with its angular direction θ.

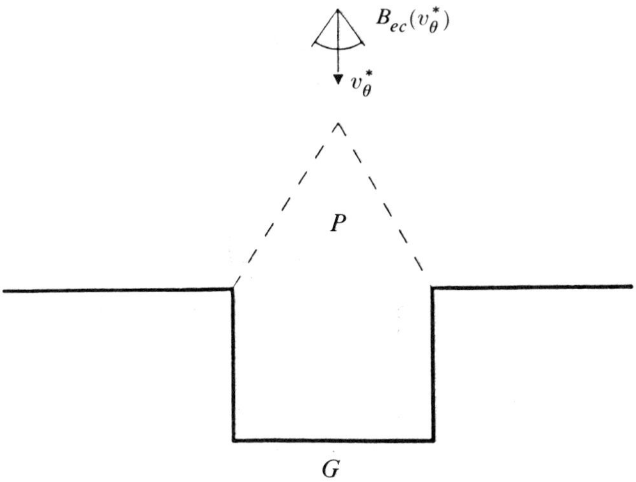

Figure 5: The goal is the region G. Sliding occurs on vertical surfaces, and sticking on horizontal ones. The commanded velocity is v_θ^*, and the control uncertainty is $B_{ec}(v_\theta^*)$. The *backprojection* of G with respect to θ is the region P.

(a) Find all vertices in the environment where sticking is *possible* under v_θ^*.

(b) At each of these vertices, erect two rays, parallel to the two edges of the *inverted* velocity cone $-B_{ec}(v_\theta^*)$.

(c) Compute the arrangement from the environment plus these additional $O(n)$ constraints.

(d) Starting at the goal edge, trace out the backprojection region.

An excellent exposition of Erdmann's algorithm can be found in [E]. With John Canny, we implemented a plane-sweep algorithm for backprojections from general polygonal goals. The idea is similar, and interested readers may find details in chapter V. While in general the complexity of these algorithms is $O((n+c)\log n)$, both methods take time $O(n \log n)$ and space $O(n)$ when the goal has $c = O(n)$ intersections with $\mathcal{P}$.

This sketch of the algorithm for verifying a compliant motion strategy gives some flavor for the kind of solution we desire. Consider the remaining problems. The trick to obtaining a computational solution to problem (1) lies in considering all possible backprojections (for all possible θ) simultaneously, and choosing θ that are suitable. In [Donald 87b,88a, Briggs 89] we show that it is possible to do this efficiently, and also provide exact algorithms for problems (3) and (4) as well.

Finally, we must note that there are many variants and facets of the guaranteed compliant motion planning problem; these we address in [Donald 87b, 88a;b, 89, 90]. For these

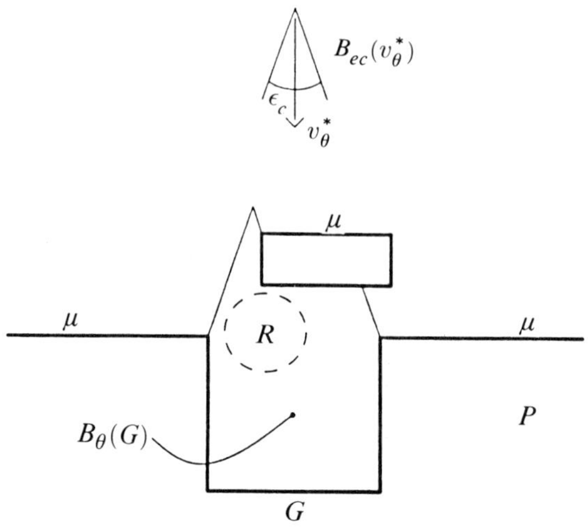

Figure 6: A compliant motion planning problem with goal G, start region R, control uncertainty ϵ_c, obstacles $\mathcal{P}$, and coefficient of friction μ. v_θ^* is a solution which is guaranteed to succeed.

problems, we also gave computational approaches.

1.8 Error Detection and Recovery

A task-level planning system requires a precise theory of Error Diagnosis and Recovery, and a method for generating sensor-based plans with built-in error detection and recovery. We call this the *Error Detection and Recovery (EDR) Theory.* I believe that the chief contribution of our work has been to define and develop a theory of EDR. Perhaps the simplest way to view EDR is as a generalization, or extension, of guaranteed geometrical planning theories. Of course, much work is required to make this notion precise, and useful; that is the bulk of the work in our monograph [Donald 89], in which in effect, we proposed a "new" theory of planning.

Our approach to EDR observes that there are certain inadequacies with the "guaranteed success" planning model. It is simply not always possible to find plans that are guaranteed to succeed. For example, if tolerancing errors render an assembly infeasible, the plan executor should stop and signal failure. In such cases the insistence on guaranteed success is too restrictive. For this reason we turn to EDR strategies. EDR plans will succeed or fail recognizably: in these more general strategies, there is no possibility that the plan will fail without the executor realizing it. The EDR framework fills a gap when guaranteed plans cannot be found or do not exist: it provides a technology for constructing plans that might work, but fail in a "reasonable" way when they cannot.

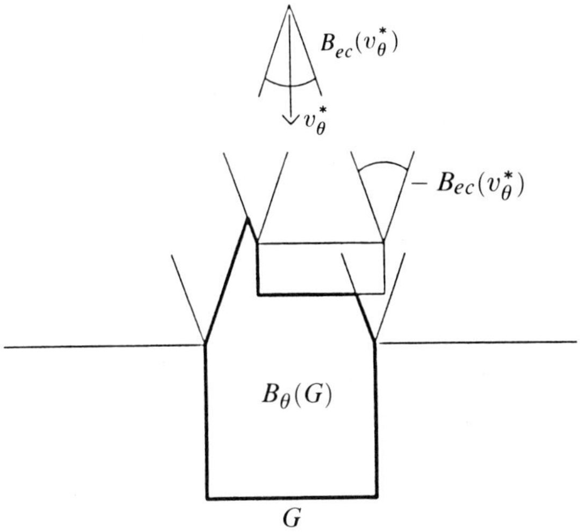

Figure 7: An illustration of Erdmann's algorithm for computing the backprojection $B_\theta(G)$ of a goal G in the plane.

The key theoretical issue is: How can we relax the restriction that plans must be guaranteed to succeed, and still retain a theory of planning that is not completely *ad hoc*? We attempt to answer this by giving a constructive definition of EDR strategies. In particular, our approach provides a formal test for verifying whether a given strategy is an EDR strategy. The test is formulated as a decision problem about projection sets in a generalized configuration space which also encodes model error. Roughly speaking, the projection sets represent all possible outcomes of a motion (the *forward projection*), and weakest preconditions for attaining a subgoal (the *preimage*).

Given the formal test for "recognizing" an EDR strategy, we can then test the definition by building a generate-and-test planner. The generator is trivial; the recognizer is an algorithmic embodiment of the formal test. It lies at the heart of this research. A second key component of the planner is a set of techniques for chaining together motions to synthesize multi-step strategies. The planner, called LIMITED, synthesizes robot control programs with built-in sensor-based EDR.

While EDR is largely motivated by the problems of uncertainty and model error, its applicability may be quite broad. EDR has been a persistent but ill-defined theme in both AI and robotics research. We give a constructive, geometric definition for EDR strategies and show how they can be computed. This theory represents what is perhaps the first systematic attack on the problem of error detection and recovery based on geometric and physical reasoning.

1.9 Experimental Results

Since 1987, my group at Cornell has implemented an experimental force-control system on our PUMA robot. In particular, we implemented an approximate generalized damper. Our PUMA 560 robot arm has a VAL-II controller, a Multibus-based 68000 computer system running NRTX (a real-time variant of UNIX) for real-time control, an Astek 6-axis force-sensing wrist, and a Monforte tool-changing gripper with finger positioning and sensing. We closed the servo loop outside of VAL-II to obtain a servo rate we estimate as approximately 20-30 hz. Probably due to the inherent passive compliance in the joints and wrist, the robot can maintain compliant contact between a (gripped) steel bolt and an aluminum surface at speeds of 1-3 inches per second.

This experimental system was sufficiently stable to perform experiments executing the plans generated by LIMITED. In particular, we executed the plans and we performed all the assemblies shown in [Donald 87b, 89], chapter I, figs. 2-13. Again, the interesting thing about these plans is that they were generated by a machine, from task-level descriptions. Thus we were able to build a small slice of a task-level system, right down to the physical robot [Donald 88c, 90; Jennings, Donald and Campbell].

The peg-in-hole plans worked very well, even with tight (1 mm) clearances. The gear-meshing plans also worked, although we found it somewhat tricky to implement the failure mode analysis "oracle" that determines when sticking or breaking contact occurs using force- and position-sensing ([Donald 87b, 88c, 89, 90; Jennings, Donald and Campbell]). For both types of plans, we found that there were unmodelled effects (impact, bouncing) that could have an effect on plan execution, and unmodelled constants (the damping constants, the amount of force to maintain for sliding on a surface, the force termination thresholds) that must be chosen by the runtime executive. A more advanced planner would take these effects and constants into account. A more sophisticated runtime force-control plan executor might adaptively adjust these parameters to a local optimum. In addition, from an experimental viewpoint, much research is needed on the implementation and choice of termination predicates.

2 Mathematical Model

2.1 Research Issues

2.1.1 Guaranteed Strategies Under Bounded Control And Sensing Uncertainty

The gross motion planning problem with no uncertainty has received a great deal of attention recently. In this problem, the state of the robot may be represented as a point in a configuration space. Thus moving from a start to a goal point may be viewed as finding an arc in free space connecting the two points. Since the robot is assumed to have perfect control and sensing, any such arc may be reliably executed once it is found. In particular, given a candidate arc, it may be tested. That is, motion along the arc may be simulated to see whether it is collision free. For example, an algebraic curve may be intersected with semi-algebraic sets defining the configuration space obstacles. In the presence of uncertainty, however, we cannot simply simulate a motion strategy to verify it. Instead, we need some technique for simulating *all* possible orbits, or evolutions of the robot system, under any possible choice of the uncertain parameters. With sensing and control uncertainty, the state of the robot must be viewed as a subset of the configuration space. Motions, then, can be viewed as mappings between these subsets. Of course there are many such subsets! From this perspective, it is clear that a chief contribution of [LMT] has been to identify and give a constructive definition for a privileged class of subsets, called *preimages*, and show that it is necessary and sufficient to search among this class. This framework appears very promising for planning guaranteed motion strategies under sensing and control uncertainty.

2.1.2 Uncertain Geometry

The [LMT] framework assumes no geometric model error. Donald [Don89] reduced the problem of planning guaranteed strategies with sensing, control, *and geometric model* uncertainty[5] to the problem of computing preimages in a (higher dimensional) generalized configuration space $C \times J$, where C is a configuration space representing the motion degrees of freedom for the robot, and J is a parameter space of model error. See sec. 2.3 for details and an example.

2.1.3 Tolerating Failure: Error Detection and Recovery

There are certain inadequacies with the planning model. The insistence that strategies be guaranteed to succeed is too restrictive in practice. To see this, observe that guaranteed strategies do not always exist. For example, in a peg-in-hole problem with model error (fig. 3) there may be no guaranteed strategy for achieving the goal, since the hole may be too small for some model error values. For these values the goal in configuration space does not exist. Because tolerances may cause gross topological changes in configuration space, this problem is particularly prevalent in the presence of model error. More generally, there may be model error values for which the goal may still exist, but it may not be reachable. For example, in a variant of the problem in fig. 6, an obstacle could block the channel to the goal. Then the goal is non-empty, but also not reachable. Finally, and most generally, there may be model error values for which the goal is reachable but not *recognizably* reachable. In this case we still cannot guarantee plans, since a planner cannot know when they have succeeded.

These problems may occur even in the absence of model error. However, without model error a guaranteed plan is often obtainable by back-chaining and adding more steps to the plan. In the presence of model error this technique frequently fails: for example, in the

[5]We use the terms model error and model uncertainty interchangeably.

peg-in-hole problem with model error, this technique will not work since no plan of any length can succeed when the hole closes up.

This is why we investigate *error detection and recovery (EDR) strategies*, as introduced in sec. 1.8. Since EDR strategies enjoy the ability to tolerate failure in a way that is robust and mathematically formal, they also provide a theoretical cornerstone for *randomized* or *probabilistic robotic strategies*, as explored by Erdmann [Erd 89] and Goldberg and Mason [GM 89].

2.2 Preimages

2.2.1 Preliminaries

We now elaborate on the definition of preimages subject to the dynamic model introduced in sec. 1.4. After the initial definitions suggested by [LMT,Ma2], this model was proposed by Erdmann [E]. Section 2.2 reviews Erdmann's definitions for the simplest termination predicates. Now, there can be much discussion as to which dynamic model is appropriate for compliant robot control, and what sensor models and termination predicates are more realistic. These are important issues. It is our goal here, to choose a *particular* model, and show how its information content can be formalized, and how, under this model, plans may be analyzed and computed. While in fact I believe that such a model is a reasonable and implementable approximation to actual robot assembly systems, I don't intend to argue that here. It is our purpose here to adopt a simple model, and then pursue the purely mathematical questions relating to reachability, sensor information, recognizability, plan computation, and complexity.

Whereas in section 1 we restricted our consideration to the configuration space $\Re^2$, we now assume our configuration space C is a Lie group, and view control velocities as generators in the Lie algebra. Sec. 1.4 may be formalized by saying that the behavior of our system is specified by a first order differential equation, called the *damper equation*, which linearly relates forces and velocities:

$$\boldsymbol{F} = \boldsymbol{B}(\boldsymbol{v} - \boldsymbol{v}_0) \tag{1}$$

Here, $\boldsymbol{F}$ is the generalized reaction force on the robot, $\boldsymbol{B}$ is the damping matrix of the form $b\boldsymbol{I}$, $\boldsymbol{v}_0$ is the control velocity, and $\boldsymbol{v}$ is the actual velocity the system.[6] One may view the vector $\boldsymbol{F}$ as the net force acting on the robot due to other objects in the robot's environment. The reaction forces due to these objects are specified by coulomb's law. $\boldsymbol{v}_0$ is related to the commanded velocity $\boldsymbol{v}_\theta^*$ as follows: $\boldsymbol{v}_0$ lies in a error ball (or cone) $B_{ec}(\boldsymbol{v}_\theta^*)$ about $\boldsymbol{v}_\theta^*$. Here is how the ball B_{ec} represents error. The possible motion directions that an unobstructed (point) robot[7] may follow when commanded to move in direction $\boldsymbol{v}_\theta^*$ lie

[6] All forces and velocities are generalized.

[7] Note it suffices to consider point robots in configuration space; see fig. 3.

within the cone $B_{ec}(\boldsymbol{v}_\theta^*)$. The effect of uncertainty on equation (1) is therefore to substitute for $\boldsymbol{v}_0$ any of the velocities in the error cone about $\boldsymbol{v}_\theta^*$.

As Erdmann [E] points out, to predict the outcome of a commanded motion $\boldsymbol{v}_\theta^*$, we must solve the generalized damper equation (1). Given that the commanded velocity in this equation (1) is subject to uncertainty

$$\boldsymbol{v}_0 \in B_{ec}(\boldsymbol{v}_\theta^*), \tag{2}$$

the solutions are actually *classes* of trajectories. Motion along configuration space surfaces corresponds exactly to motion subject to a holonomic constraint: the damper equation (1) must be solved with contact for the surface in question. See fig. 11.

We will assume that

1. The control velocity $\boldsymbol{v}_0$ can vary arbitrarily within $B_{ec}(\boldsymbol{v}_\theta^*)$.

2. Trajectories must be the result of integrating the first order damper equation (1):

$$\boldsymbol{p}(t_0) = \boldsymbol{p}(0) + \int_0^{t_0} \boldsymbol{v}(t)dt. \tag{3}$$

Definition 2.1 *A trajectory that* satisfies the damper equation with uncertainty relative to a commanded velocity $\boldsymbol{v}_\theta^*$ *is a mapping*

$$\begin{array}{rcl} T : [0,\infty) & \longrightarrow & TC \\ t & \mapsto & (T_p(t), T_v(t)) \end{array} \tag{4}$$

such that at all times t there exists a $\boldsymbol{v}_0(t)$ satisfying (2) with

$$\boldsymbol{F}(t) = \boldsymbol{B}(\boldsymbol{v}(t) - \boldsymbol{v}_0(t)), \tag{5}$$

and (3) relates T_p and T_v.

Definition 2.2 *Define $\mathcal{T}(\boldsymbol{v}_\theta^*)$ to be the set of all trajectories that satisfy the damper equation with uncertainty relative to $\boldsymbol{v}_\theta^*$.*

2.2.2 Sensing Model

Position sensing is specified as follows. For all times t, the sensed position $\boldsymbol{p}^*$ must lie within a ball $B_{ep}(\boldsymbol{p})$ about the actual position $\boldsymbol{p}$.

Now, given the damper equation (1), velocity- and force-sensing are equivalent. Hence we assume that for all times t, the sensed velocity $\boldsymbol{v}^*$ must lie within a ball $B_{ev}(\boldsymbol{v})$ about the actual position $\boldsymbol{v}$.[8]

[8]For technical reasons, we must also assume symmetry, i.e., that $\boldsymbol{p} \in B_{ep}(\boldsymbol{p}^*)$ and $\boldsymbol{v} \in B_{ev}(\boldsymbol{v}^*)$ as well.

2.2.3 Termination Predicate and Forward Projections

When the motion is initiated, the termination predicate knows the commanded velocity $\boldsymbol{v}_\theta^*$, the error bounds B_{ec}, B_{ep} and B_{ev}, and the set of goals to be reached, $\{ G_\alpha \}$. In addition, it also knows a bound on the initial conditions; in other words, it knows an initial set R such that the resulting trajectory must originate in R. In this survey, we do not assume that the termination predicate can "remember" past sensor values, although this model has been considered by Mason [Ma2] and Erdmann [E].

We assume that our termination predicate continuously monitors sensing values and time, and that it terminates the motion when the trajectory is guaranteed to have entered a goal. At that point the termination predicate returns which goal $G \in \{ G_\alpha \}$ has been achieved. Hence, we view the computational role of the termination predicate as follows: at all times, the termination predicate computes the set of effective interpretations of its sensors, and checks to see whether this set is entirely contained within some goal. When containment is decided, the predicate returns.

In order to model this process mathematically, it is useful to speak of a trajectory as being *consistent* with a sensor value—or the other way around. Hence:

Definition 2.3 *A trajectory T is* sensor consistent at time t with reading $(\boldsymbol{p}^*, \boldsymbol{v}^*)$ *when*

$$(T_p(t), T_v(t)) \in B_{ep}(\boldsymbol{p}^*) \times B_{ev}(\boldsymbol{v}^*). \tag{6}$$

Since the termination predicate knows the controls, and an initial set R in which the motion is guaranteed to originate, it can construct a bound on all possible reachable states. This bound is called the *forward projection.*

Definition 2.4 *The* forward projection at time t of a set of initial conditions R, under commanded velocity $\boldsymbol{v}_\theta^*$ subject to uncertainty (2), *is given by*

$$F_\theta(R,t) = \{ T(t) \mid T \in \mathcal{T}(\boldsymbol{v}_\theta^*) \text{ and } T_p(0) \in R \}. \tag{7}$$

Correspondingly,

Definition 2.5 *The* timeless forward projection *is*

$$F_\theta(R) = \bigcup_{t \geq 0} F_\theta(R,t). \tag{8}$$

2.2.4 Application: Computing Forward Projections

We now consider the computation of forward projections in various simple configuration spaces. Note that forward projections lie in phase space. Now, for a configuration space C, let $\pi : TC \to C$ be the canonical projection map. Buckley [Buc] developed the first algorithms in $C = \Re^3$ for computing the position component of the forward projection $\pi F_\theta(R)$. By the results of Canny and Reif [CR], these sets can have exponential size, and the following decision problem is computationally intractable:

Lemma 2.1 [CR]. *In the configuration space $\Re^3$, given a polyhedral start region R, a polyhedral environment $\mathcal{P}$ of n vertices, control uncertainty ϵ_c, coefficient of friction μ, and a polyhedral goal G, and a commanded motion $\boldsymbol{v}_\theta^*$, deciding containment of a point $\boldsymbol{x}$ in the forward projection $\pi F_\theta(R)$ is $\mathcal{NP}$-hard.*

This immediately implies the following

Corollary 2.2 [CR]. *The One-Step Compliant Motion Verification Problem (problem 2 in sec. 1.5) $(R, \mathcal{P}, \epsilon_c, \mu, G)$ is $\mathcal{NP}$-hard in $\Re^3$.*

However, the planar case turns out to be tractable. By generalizing results of Erdmann [E], Donald and Canny developed forward projection algorithms for that are optimal $(O(n \log n))$ in the plane [Don 89].

We now consider some computational complications of introducing rotations. Note that our discussion generalizes straightforwardly to the configuration space $\Re^3 \times SO(3)$ of rigid body euclidean motions. For the remainder of section 2.2.4, it it convenient to drop the boldface notation for points and tangent vectors to C.

In section 1 we considered the configuration space $\Re^2$ to develop intuition. Suppose we continue to assume the obstacles and peg are polyhedral, but we allow the peg to rotate as well as translate. This lifts the problem into the three dimensional manifold corresponding to the special euclidean group acting on the plane, which we will represent as $C = \Re^2 \times S^1$. This configuration space may be given coordinates (x, y, θ), or, if we desire an "algebraic" configuration space, we may use the coordinates (x, y, u) where $u = \tan\frac{\theta}{2}$. Configuration space obstacles in this space are 3-dimensional semi-algebraic sets (see figs. 8-10, from Brost's work [Brost 89]).[9] These obstacles are generated by one triangular moving "peg" A, and one triangular stationary "obstacle" B. The surfaces bounding these obstacles are ruled, algebraic surfaces; they are simultaneously linear in x and y, and quadratic in u. Each configuration space surface S, or "C-surface" in figs. 8-10 corresponds exactly to a particular holonomic constraint. In particular, each C-surface is "generated" by a particular contact type. That is, each C-surface is generated by the constraint that a vertex of A touch an edge of B, or that an edge of A touch a vertex of B. Note that $C = \Re^2 \times S^1$ is made into a Riemannian manifold by the choice of inner product as follows: If v is a tangent vector at $x \in C$, then $\langle v, v \rangle_x = v \boldsymbol{I}_x v$ where $\boldsymbol{I}_x$ is the moment of inertia tensor at x. This means that $\langle \cdot, \cdot \rangle_x$ is a quadratic form that computes the energy of the system, and hence is a "natural" choice in the sense of Lagrangian dynamics. Each C-surface S

[9]These figures are very similar to those in [Brost 89] but were kindly provided to me by the author and are reprinted with permission here.

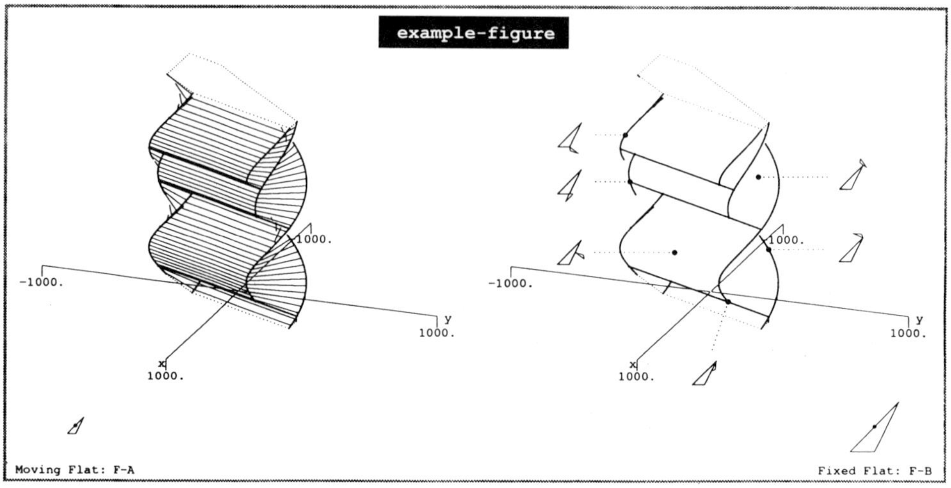

Figure 8: In the configuration space $\Re^2 \times S^1$, shows the configuration space obstacle for a small triangular "peg" A and a large triangular "obstacle" B. This figure shows how each C-surface bounding the obstacle is generated by a contact pair between a vertex (resp. edge) of A and an edge (resp. vertex) of B. Figs. 8-10 reprinted with permission from [Brost 89], and from new figures provided by R. Brost for this article.

hence inherits this inner product. We write the normal to S at x under this inner product as N_x. N_x corresponds to the direction of the pure constraint force for S satisfying the D'Alembert-Lagrange principle.

In free space (the complement of the obstacles), the set of all possible motions simply fans out in a 3D cone. The dynamics of sliding on surfaces, even under generalized damper dynamics, is more complicated. In particular, the dynamics changes from point to point as a s.a. function of orientation. See fig. 11, which shows a typical C-surface S. The dynamics form a differential constraint at each point x on S. The differential constraint is a projection of the 3D velocity uncertainty cone $B_{ec}(v_\theta^*)$ onto the tangent space to S at x. This projection forms a 2D cone; the differential constraint specifies that for a trajectory satisfying the damper equation with uncertainty relative to v_θ^*, and subject to constraint S, that the velocity $T_v(t)$ at $x = T_p(t)$ must lie inside the projected cone. The projection mapping, and hence the cone, varies from point to point. Now, this projection mapping is given by the mechanics of coulomb friction; since friction changes in a non-linear way

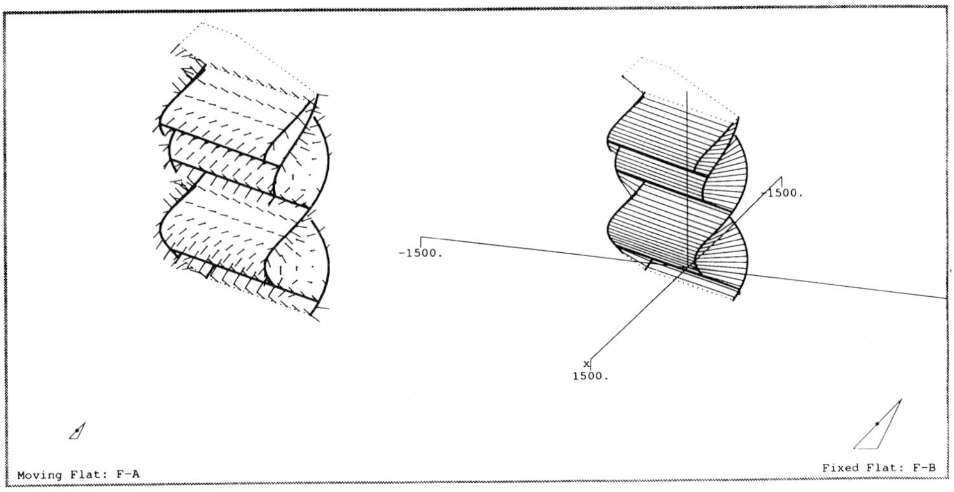

Figure 9: Normals to the C-surfaces. Each C-surface S is ruled and algebraic, and is simultaneously linear in x and y, and quadratic in $u = \tan\frac{\theta}{2}$.

over the surface,[10] the projection does also. Erdmann [E] gives the form of the projection map π_x, which is semi-algebraic (s.a.). We note that computing forward projections in $\Re^2 \times S^1$ has both a "dynamical systems" component (finding the outer envelope of all possible evolutions of the dynamical system π_x, subject to a holonomic constraint S), and a "combinatorial" component (there are a finite number of C-surfaces S that we can slide on).

Example: We derive the map π_x for the case of no friction. C is a parallelizable n-manifold, hence $TC \cong C \times \Re^n$. Let $\phi_x : T_eC \rightarrow T_xC$ smoothly identify each tangent space with the tangent space at the identity. Now, we may view $x \in S$ as a point in C, and hence identify T_xS with a subspace of T_xC, such that the normal N_x is orthogonal to T_xS. Now, define $p_x : T_xC \rightarrow T_xS$ to be the projection of T_xC onto T_xS along N_x. Then it is easily seen that $\pi_x = p_x \circ \phi_x$. With friction, the situation is somewhat more complicated, but the definition of π_x has the same spirit. See Erdmann [E].

Definition 2.6 *Let S be an algebraic surface in the configuration space C, and $R \subset S$ a set of initial conditions. Let $B_{ec}(v_\theta^*)$ be the control uncertainty cone in the tangent space*

[10]That is, the damper equation (1) during contact is non-linear.

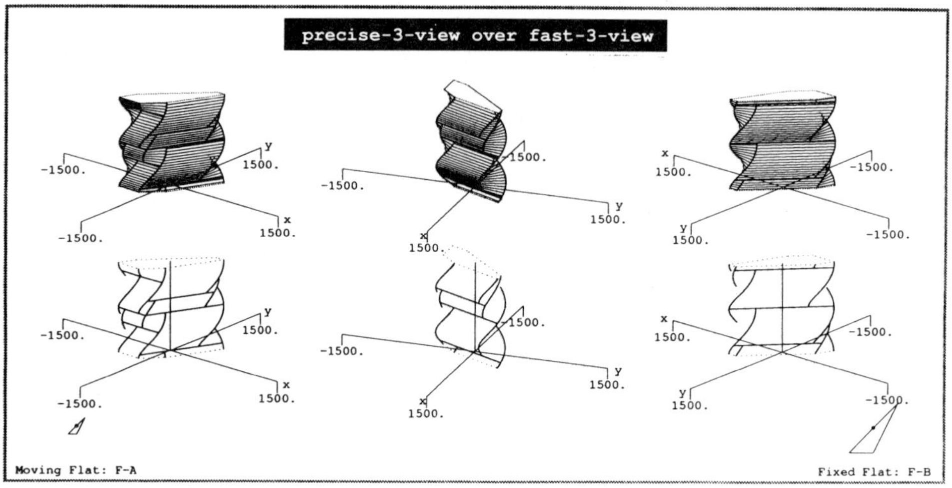

Figure 10: Three different "views" of the configuration space obstacles.

T_eC to C at the identity. Let $\pi_x : T_eC \to T_xS$ be a projection map that varies with x and is semi-algebraic. Then the forward projection $F_\theta(R)|_S$ of R under θ subject to S *is defined as follows:*

$F_\theta(R)|_S$ is the image of all trajectories T with $T_p(0) \in R$, such that at $T(t) = (x, v)$, with controls $v_0(t)$ satisfying (2), one of the following holds:

1. (Maintain contact): $\langle v_0(t), N_x \rangle_x \leq 0$ *and* $v \in \pi_x(B_{ec}(v_\theta^*))$, or

2. (Break Contact): $\langle v_0(t), N_x \rangle_x \geq 0$ *and* $v = v_0(t)$.

The forward projection with time subject to S is defined analogously. We can now state

Problem 5. Suppose S is a ruled, quadric C-surface in the configuration space $\Re^2 \times S^1$, as described above. Find combinatorially precise algorithms for computing $F_\theta(R)|_S$, and for deciding containment of a point in this forward projection.

At this time, problem (5) is open, although Erdmann [E] and Brost [Brost 85] have given "numerical" approximation algorithms for backprojections, and Donald [Don 89] has

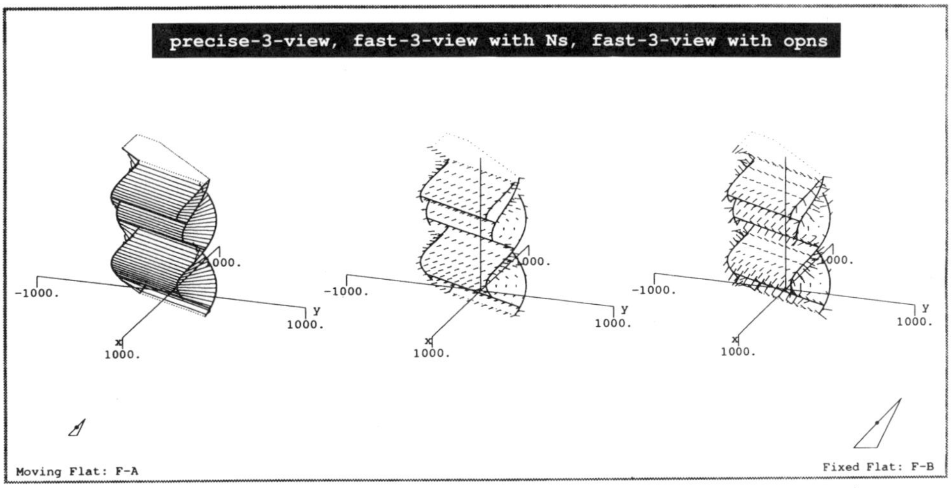
precise-3-view, fast-3-view with Ns, fast-3-view with opns
-1000.
1000.
y
x
1000.
Moving Flat: F-A
Fixed Flat: F-B

Figure 10a

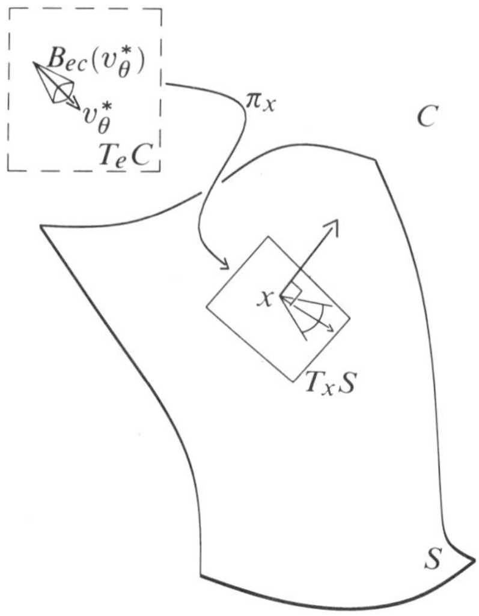

Figure 11: S is an algebraic surface in the lie group C. π_x projects control set B_{ec} from the Lie algebra of C onto the tangent space to S at x. π_x is a semi-algebraic map, (and s.a. parameterized by x). The projected velocity cone is a differential constraint that the velocity of admissible trajectories in TS must obey.

given (similar) "numerical" algorithms for forward projections. In a recent breakthrough, Canny has developed a combinatorially precise algorithm for computing fully-general LMT multi-step strategies in the configuration space $\Re^3$, using the termination predicate in eq. (10). No such algorithms exist yet for $\Re^2 \times S^1$ or $\Re^3 \times SO(3)$, although there are approximation algorithms; some of these apply to restricted subproblems and are precise algorithms [Don88a, Briggs, FHS].

2.2.5 The Role of the Forward Projection in Preimages

Now, knowing where a motion started gives the termination predicate additional constraint on the set of effective sensor interpretations. For example, suppose that the position sensors read $\boldsymbol{p}^*$. *A priori*, this means that the actual position $\boldsymbol{p}$ can lie anywhere in the ball $B_{ep}(\boldsymbol{p}^*)$. However, suppose that the termination predicate has computed the forward projection $\pi F_\theta(R)$; this is an "upper bound" on all reachable positions. If part of the sensor ball $B_{ep}(\boldsymbol{p}^*)$ lies outside the forward projection, then those configurations are clearly excluded as effective sensor interpretations, since they are unreachable. More specifically, we have that

Proposition 2.3 *The effective interpretation of the sensor values* $(\boldsymbol{p}^*, \boldsymbol{v}^*)$ *at time* t *is*

$$F_\theta(R,t) \cap B_{ep}(\boldsymbol{p}^*) \times B_{ev}(\boldsymbol{v}^*). \qquad (9)$$

2.2.6 Definition of Preimages

We can now define preimages. First, we define the "set of successful controls" $S(R, \{ G_\alpha \})$:

Definition 2.7 *For a start region R and a set of goals $\{ G_\alpha \}$, Define the set $S(R, \{ G_\alpha \})$ as follows. $\boldsymbol{v}_\theta^* \in S(R, \{ G_\alpha \})$ if and only if:*

For all trajectories $T \in \mathcal{T}(\boldsymbol{v}_\theta^*)$ consistent with the damper equation under uncertainty, there exists a time $t \geq 0$ such that for all sensor values $(\boldsymbol{p}^*, \boldsymbol{v}^*)$ consistent with $T(t)$, there exists some goal $G \in \{ G_\alpha \}$, such that $\pi^{-1}G$ contains the set of effective sensor interpretations (9), that is,

$$F_\theta(R, t) \cap B_{ep}(\boldsymbol{p}^*) \times B_{ev}(\boldsymbol{v}^*) \subset \pi^{-1}G. \tag{10}$$

Then preimages are as follows:

Definition 2.8 *The* Preimage of a set of goals $\{ G_\alpha \}$ with respect to a start region R, for a commanded velocity $\boldsymbol{v}_\theta^*$, *is defined as*

$$P_{R,\theta}(\{ G_\alpha \}) = \{ \boldsymbol{p} \in R \mid \boldsymbol{v}_\theta^* \in S(R, \{ G_\alpha \}) \}. \tag{11}$$

Lozano-Pérez, Mason, and Taylor [LMT] envisioned a back-chaining planner that would recursively compute preimages of a goal until the start region was contained in a preimage. For example, starting with a goal G_0, one first computes the preimage G_1 of G under θ_1. Then one computes the preimage G_2 of G_1 under θ_2, and so forth, until the start region R is contained in G_n. Then one executes the plan $(\theta_n, \ldots, \theta_1)$, and the termination predicate (10) is guaranteed to recognize entry into each successive subgoal G_i.[11]

One then asks, what characterizes preimage sets, and when are they suitable subgoals for the next level of backchaining? The first result is:

Theorem 2.4 *(Fixed-Point Theorem [LMT]). In order that $P_{R,\theta}(\{ G_\alpha \})$ be a suitable subgoal for the next level of backchaining, the equation*

$$R = P_{R,\theta}(\{ G_\alpha \}) \tag{12}$$

must hold.

Hence, we may view the problem of guaranteed motion planning with uncertainty as solving the preimage equation (12) for R. The preimage equation is a consequence of the following property of the preimage mapping:

Lemma 2.5 (Erdmann). *The mapping $P_{R,\theta}$ is idempotent, when viewed as a function of R, for a fixed collection of goals.*

[11]This example does not illustrate branching, or conditional plans.

Comment: (On distinguishable goal sets). Consider the following preimage equation:

$$P_{R,\theta}(\{\, G, H \,\}) = R. \tag{13}$$

We say that the preimage (13) is taken *with respect to* R. (13) means that the preimage of the set of goals $\{\, G, H \,\}$, with respect to commanded velocity v_θ^*, is all of R. Note that by def. 2.7, when we have a *set* of goals, the termination predicate must return *which* goal (G or H) has been achieved. This is different from $P_{\theta,R}(G \cup H)$, which means the termination predicate will halt saying "we've terminated in G or H, but I don't know which." The region R appears on both sides of (13) because the preimage depends on knowing where the motion started. Thus solving preimage equations like (13) for R is like finding the fixed point of a recursive equation.

2.2.7 Preimage Approximation Theory

Preimages exhibit several undesirable qualities that make for computational difficulties. First, in general, maximal preimages are not unique. Second, in some cases, maximal preimages do not even exist. This led Erdmann to pursue an agenda that we may term "preimage approximation theory", in which he attempted to find a preimage-like mathematical object that enjoyed the properties of maximality, uniqueness, and efficient computability. This object he termed the *backprojection.*

Example: Fig. 12 illustrates the difference between backprojections and preimages. Here the radius of position sensing uncertainty is greater than twice the diameter of the hole. Sliding occurs on all surfaces. Furthermore, we assume that the robot has no sense of time (i.e., no clock)—for example, it might be equipped with independent x and y contact sensors that only fire once each.

We formalize this sensor model by assuming that the robot has accurate force-sensing in the y-direction, and infinitely bad force-sensing in the x-direction, and that there is no friction. The start region R can be all of free space. The backprojection $B_\theta(G)$ strictly contains the preimage $P_{R,\theta}(G)$: while all points in the backprojection are guaranteed to reach G, the sensing inaccuracy is so large that the termination predicate cannot tell whether the goal or the left horizontal surface has been reached. Only from the preimage can entry into G be recognized.

We may define the *simple backprojection* $\overline{P_{R,\theta}}$ of a set of goals to be the preimage with perfect position- and velocity- sensing. We observe that every preimage is contained in a simple backprojection. Hence, backprojections only address goal *reachability*, whereas preimages capture reachability *and recognizability*. No notion of termination predicate is needed to formalize the notion of backprojection, although in our case, it is useful pedagogically to think of a termination predicate with "perfect" sensing. Erdmann [E] made precise the notion that "backprojections may be used to approximate preimages." The idea

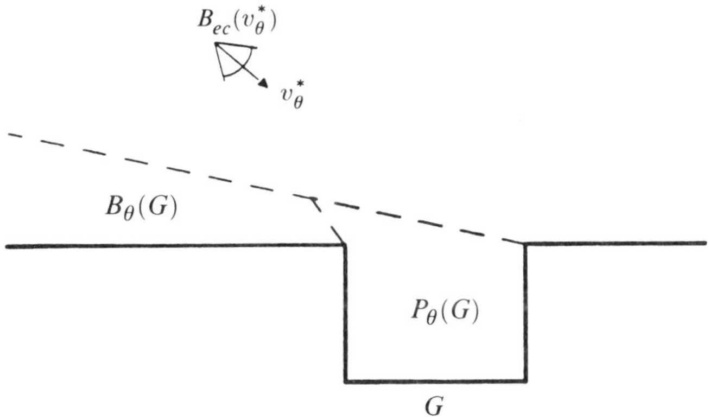

Figure 12: Here, the radius of the position sensing uncertainty ball is twice the width of the hole. Sliding occurs on all surfaces under the control velocities shown. Let the start region R be all of free space. The preimage of the goal under commanded velocity v_θ^* is $P_\theta(G) = P_{R,\theta}(G)$. The backprojection $B_\theta(G)$ strictly contains this preimage: while all points in the backprojection are guaranteed to reach G, the sensing inaccuracy is so large that the termination predicate cannot tell whether the goal or the left horizontal surface has been reached. Only from the preimage can entry into G be recognized.

is as follows. First, Erdmann observed that whether or not a point is in a backprojection depends solely on the properties of the point, and not on the properties of neighboring points. That led him to define a "maximal" backprojection $\overline{M_\theta}$ as the union of "simple" backprojections:

$$\overline{M_\theta} = \bigcup_R \overline{P_{R,\theta}}(\{ G_\alpha \}). \tag{14}$$

Note that by idempotency, it suffices to consider only sets R such that $R = \overline{P_{R,\theta}}(\{ G_\alpha \})$. It is clear, therefore, that the union of simple backprojections is also a simple backprojection:

$$\overline{M_\theta} = \overline{P_{\overline{M_\theta},\theta}}(\{ G_\alpha \}). \tag{15}$$

Hence, the absence of termination predicates (or, equivalently, the assumption of perfect termination predicates) makes it possible to define a unique maximal set $\overline{M_\theta}$ such that under controls v_θ^*, from $\overline{M_\theta}$ we are guaranteed to reach a goal $G \in \{ G_\alpha \}$. Erdmann called the mapping $\theta \times \{ G_\alpha \} \mapsto \overline{M_\theta}$ the *maximal backprojection*, denoted

$$B_\theta(\{ G_\alpha \}) = \overline{M_\theta}. \tag{16}$$

Hence we have

Lemma 2.6 *For all R, θ, and $\{ G_\alpha \}$,*

$$P_{R,\theta}(\{ G_\alpha \}) \subset \overline{P_{R,\theta}}(\{ G_\alpha \}) \subset B_\theta(\{ G_\alpha \}) = B_\theta\Big(\cup_\alpha \{ G_\alpha \}\Big). \tag{17}$$

2.2.8 Erdmann's Structure Equation

Now, Erdmann showed that every preimage $R = P_{R,\theta}(\{ G_\alpha \})$ may be "extended" to an "almost simple" preimage $A(R)$, where $A(R)$ contains R and has the general form

$$A(R) = \pi F_\theta(R) \cap B_\theta(E(R)), \tag{18}$$

where $E(R)$ is a subset of the union of the goals $\cup_\alpha \{ G_\alpha \}$. This says that given R, θ, and $\{ G_\alpha \}$, one may, in principle, "shrink" the goals to compute $E(R)$, and then compute $A(R)$ as in (18).

It remains to define $E(R)$, which Erdmann called the *First Entry Set*.

Definition 2.9 *For a trajectory T, let $\mathcal{S}_T$ be the set of successful termination times for R, θ, and $\{ G_\alpha \}$. That is, $\mathcal{S}_T$ is the set of all times $t \geq 0$ such that for all sensor values $(\mathbf{p}^*, \mathbf{v}^*)$ consistent with $T(t)$, there exists some goal $G \in \{ G_\alpha \}$, such that $\pi^{-1}G$ contains the set of effective sensor interpretations (eq. (9)).*

Definition 2.10 *Suppose* $R = P_{R,\theta}(\{ G_\alpha \})$ *is preimage. Define the* First Entry Set $E(R)$ *of* R *to be the set of all "first entry points" of trajectories that start in* R*. That is,*

$$E(R) = \{ T_p(t_0) \mid T \in \mathcal{T}(v_\theta^*), T_p(0) \in R, \text{ and } t_0 = \inf(\mathcal{S}_T) \}. \tag{19}$$

Hence, we have that every suitable preimage R is a subset of an "almost simple" preimage" $A(R)$, and $A(R)$ is also a suitable preimage. We also have that $A(R)$, the extension of R, is "maximal" in the sense that A is idempotent. We call eq. (18) the *structure equation* for preimages and backprojections. On a practical note, while computing first entry sets may be in general, as hard as computing preimages, the utility of this definition is that many, easy-to-compute goal-shrinking schemes can quickly be proved correct. That is, a conservative goal-shrinking scheme is correct if it computes a subset G of $E(R)$. Then we may compute a preimage using the structure equation.

2.3 Multi-Step Strategies

Having reduced motion planning under uncertainty to essentially "preimage-theoretic" equations, motion strategies can be synthesized by solving these preimage equations. We now present a worked-out example of a motion plan using preimages. The motion problem is grasp-centering for a robot gripper in the presence of model error. A guaranteed plan is found by solving the preimage equations. The example[12] illustrates the use of the preimage framework to derive a multi-step motion strategy in the presence of model error. The strategy employs time-sensing and force-sensing.

Consider the grasp-centering problem shown in fig. 13. The task is to center the robot gripper over the block D. The gripper can translate but not rotate in the plane. In its start position, the gripper is somewhere over D, such that the bottom of the fingers FA and FB are below the top of D. The width of D is unknown, but must be less than the distance between FA and FB. We assume D is fixed (it cannot be accidentally pushed).

Hence we can regard this as a planning problem with model error. C is taken to be the cartesian plane, and J, the space of model error, is a bounded interval of the positive reals. Our first question is, what does the generalized configuration space $C \times J$ look like? This is easily answered by considering the motion planning problem in fig. 14. The problem is to find a motion strategy for a point robot so that it can achieve a goal exactly halfway between the blocks A and B. The distance α between A and B is unknown and positive. The point robot is known to start between A and B. Again, the point can translate in the plane. The distance α is the model error parameter. It is easy to see that the problems in figs. 13 and 14 are equivalent. The generalized configuration space for fig. 14 is, in fact, the same as that depicted in fig. 18, as we will see in sec. 2.4.

[12]This problem arose in discussions with Tomás Lozano-Pérez, John Canny, and Mike Erdmann. It appears in [Don89].

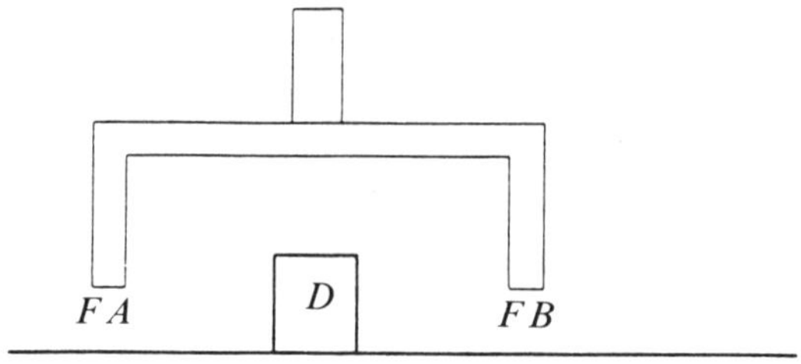

Figure 13: The grasp centering problem. The width of the block D on the table, and the position of the gripper are only known approximately.

Next, we assume that the robot has perfect control, perfect velocity sensing, and a perfectly accurate sense of time. However, it has infinite position sensing error. [13]

Now, since the gripper starts over D with the bottom of the fingers below the top of D, and since the robot has perfect control, it suffices to consider the x axis of C. Since the y axis can be ignored, we develop our example in the plane, that is, in the generalized configuration space where C and J are both one-dimensional. This 2D generalized configuration space is shown in fig. 15, which is essentially an x-J cross-section of the full 3D generalized configuration space $\Re^2 \times J$ shown in fig. 18, holding y constant with α constrained to be positive. In fig. 15, L and R are left and right obstacle edge boundaries generated by A and B. The goal is the line in free-space bisecting L and R. The start region T is the triangular region in free-space between L and R. (T is the convex hull of L and R).

Now, since motion across J is not permitted, all motions are parallel to the x axis, that is to say, horizontal in fig. 15. There are only two kinds of motions the planner can command. Let $+$ denote a motion to the right, and $-$ a motion to the left. We assume the robot has perfect control over the magnitude as well as the direction of the commanded velocity.

See fig. 15. Now, if α is a point on the J axis, let E_α be the point on the left obstacle edge L with J coordinate α. We will denote the collection of all such points on L by $\{ E_\alpha \}$. Let S_α denote the maximal line segment within T containing E_α and parallel to

[13]This example is easily generalized to non-zero control, time-sensing, and force-sensing error, and finite position-sensing error. This requires giving the goal non-empty interior, however.

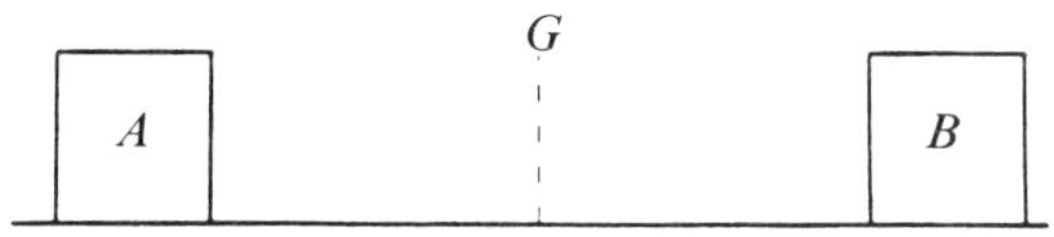

Figure 14: An equivalent problem. A point robot must be navigated halfway between the blocks A and B. The distance between A and B is not known. The robot has force sensing, and a clock. However, it has poor position sensing. We regard C as $\Re^2$ and J, the space of model error, as the bounded interval $(0, d]$ for d positive. The generalized configuration space for this problem is $\Re^2 \times J$, and is shown in fig. 18.

G. Formally, if E_α has coordinates (x, α), then S_α is the line segment extending from E_α to (x, d) where d is an upper bound on the distance between A and B. We denote the collection of all lines S_α by $\{ S_\alpha \}$.

At this point we are prepared to derive a motion strategy for centering the grasp, that is, for attaining G from T. The strategy has three steps. The termination conditions for the motions involve time- and force-sensing. Here is the motion strategy in qualitative terms:

Strategy Guarantee-Center

1. *Command a motion to the right. Terminate on the right edge R based on force sensing.*
2. *Command a velocity of known magnitude to the left. Terminate when in contact with the left edge L, using force sensing. Measure the elapsed time of the motion. Compute the distance traversed. This gives exact knowledge of where the motion terminated on L. The effect of this step is to measure the distance α between the blocks.*
3. *Move distance $\frac{\alpha}{2}$ to the right, terminating in G based on time sensing.*

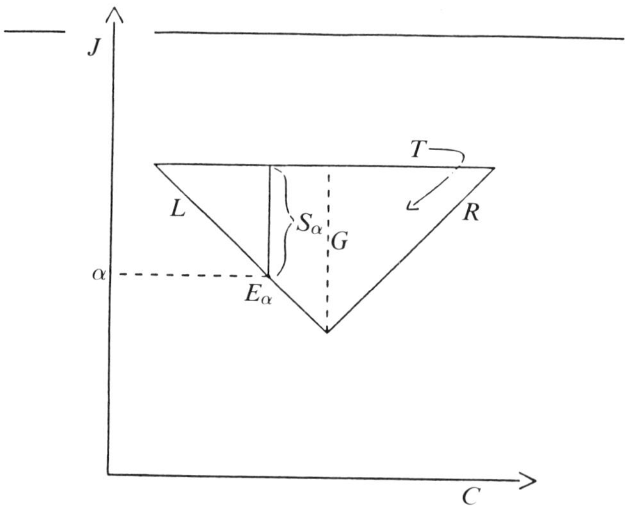

Figure 15: Assuming that the gripper fingers are initially lower than the top of the block D, the y dimension can effectively be ignored. This allows us to examine an x-J cross-section of the 3D generalized configuration space $\Re^2 \times J$ in fig. 18. We treat C as the x axis of motion freedom, yielding a 2D $C \times J$ planning space. L and R are obstacle boundaries in generalized configuration space . The goal is the bisector G between L and R in free-space. The start region T is the triangular region between L and R. E_α is a point on L. S_α is a line in T parallel to G and containing S_α.

We now derive this strategy by solving the preimage equations for the motion planning problem.

First, note that if the run-time executive knows that the robot is inside a *particular* S_α, then G can be reliably achieved by commanding a motion to the right. Since the robot has perfect control and time sensing, the motion can be terminated after moving distance $\frac{\alpha}{2}$, that is, exactly when the line G is achieved. Using the preimage notation, we write this as

$$P_{+,\{S_\alpha\}}(G) = \{S_\alpha\}. \tag{1}$$

Next, we take the collection $\{S_\alpha\}$ as a set of subgoals, and try to find a motion that can recognizably attain this collection, and, furthermore, can distinguish which S_α the motion achieves. Consider a leftward motion starting from anywhere on the right edge R. The robot does not know where on R the motion starts, however. To recognizably achieve some S_α, such a motion should move leftward, and terminate when force-sensing indicates that L has been reached. If the termination predicate measures the elapsed time of the motion, and knows the magnitude of the commanded velocity, then it can recognize which point E_α has been reached, and hence which subgoal S_α has been achieved. Writing this down in preimage equations,

$$P_{-,R}(\{S_\alpha\}) = P_{-,R}(\{E_\alpha\}) = R. \tag{2}$$

Finally, the right edge R may be achieved from anywhere within the start region T by moving rightward, and terminating when force sensing indicates contact. This is simply

$$P_{+,T}(R) = T. \tag{3}$$

It is instructive to examine the termination conditions for motions (1)–(3). In developing the LMT framework for planning guaranteed strategies, [Erdmann] developed an elegant formalization of the question, "Using sensors and history, when can the termination predicate decide that a motion has recognizably entered a goal G_β?" The answer was as follows. Assume that G_β has been lifted into phase-space. Let R be the start region. The forward projection, $F_\theta(R)$ captures the notion of history: it is all positions and velocities that can be reached given that the motion started in R. At a particular instant t in time, let $B_{ep}(t)$ and $B_{ev}(t)$ be the sets of possible positions and velocities. These are the sensing uncertainty balls about a sensed position and velocity in phase space at time t. Thus sensing provides the information that the actual position and velocity must lie within the set $B_{ep}(t) \times B_{ev}(t)$. The forward projection further constrains the actual position and velocity to lie within $F_\theta(R)$. Thus the termination predicate can terminate the motion as having recognizably reached G_β when

$$F_\theta(R) \cap (B_{ep}(t) \times B_{ev}(t)) \subset G_\beta. \tag{$*$}$$

Now, let $F_\theta(R,t)$ denote the *time-indexed* or *instantaneous forward projection* of R under θ at time t. $F_\theta(R,t)$ denotes the set of positions and velocities that are possibly achievable at elapsed time t, under motion θ, given that the motion started in R. The termination predicate in this case monitors a clock, in addition to position and velocity sensors. In motion (1), only the time-indexed forward projection $F_+(S_\alpha, t)$ is relevant to deciding termination. The motion terminates when $F_+(S_\alpha, t) \subset G$. Motion (3) can be terminated using pure force sensing. It could also be terminated using time, since there exists some t for which $F_+(T, t) = R$. In motion (2), both force sensing and time are required to terminate within a distinguishable E_α. The general form of the termination condition for all three cases is as follows. The termination predicate has the form

$$F_\theta(U, t) \cap (B_{ep}(t) \times B_{ev}(t)) \subset G_\beta$$

for a goal G_β and a start region U. (Assume that all subgoals have been lifted into phase space). In our case, position sensing error is infinite, so $B_{ep}(t)$ is $C \times J$. Let us denote $(C \times J) \times B_{ev}(t)$ by the simpler expression $B_v(t)$. Then the termination conditions for motions (1)–(3) are as follows. For the first motion (3), to terminate, we must have

$$F_+(T, t) \cap B_v(t) \subset R. \tag{4}$$

For the second motion (2) to terminate, we must have

$$F_-(R, t) \cap B_v(t) \subset S_\alpha \tag{5}$$

for some S_α. We think of the termination predicate as "returning" this S_α. Finally, for termination of the last motion (1), we must have

$$F_+(S_\alpha) \cap B_v(t) \subset G, \tag{6}$$

where the S_α in (6) is the same as the one returned by the termination predicate after the second motion as the satisfying assignment for (5).

Finally, note that time is the source of some complexity in this example. This complexity might be removed by employing a distance sensor instead. The output of such a sensor could be modeled as position sensing in J. The sensing action in J would entail measuring the distance between A and B. This relaxes the assumption of no position sensing in the J dimensions, but such modification to the generalized configuration space framework is trivial. With this modification, B_{ep} is simply regarded as a product of a position sensing ball in C and a position sensing set in J.

This concludes the example. We have shown how to derive a multi-step guaranteed motion strategy in the presence of model error. The strategy was derived by solving the preimage equations in generalized configuration space for the motion plan. These preimage equations made the role of time- and force-sensing explicit in deriving conditions for distinguishable termination in a collection of subgoals.

2.4 Representing Model Error

The [LMT] framework assumes no geometric model error. Donald [Don89] reduced the problem of planning guaranteed strategies with sensing, control, *and geometric model* uncertainty to the problem of computing preimages in a (higher dimensional) generalized configuration space, which encodes model error as a kind of nonholonomic constraint. In fact, we used this method implicitly in the previous section 2.3. We new review the reduction by carefully examining some very simple planning problems with model error. These problems in fact motivate the definition of *Error Detection and Recovery (EDR) strategies.* Of course, this does not mean that EDR is limited to situations with model error.

2.4.1 A Simple Example: The Variable-Width Peg-In-Hole

Example 1: Consider fig. 16. There is position sensing uncertainty, so that the start position of the robot is only known to lie within some ball in the plane. The goal is to bring the robot in contact with the right vertical surface of A.

We will simplify the problem so that the computational task is in configuration space. This transformation reduces the planning task for a complicated moving object to navigating a point in configuration space. Consider fig. 17. The configuration point starts out in the region R, which is the position sensing uncertainty ball B_{ep} about some initial sensed position. As usual, to model sliding behavior, we will assume Coulomb friction and generalized damper dynamics, which allows an identification of forces and velocities. Thus the commanded velocity v_0 is related to the effective velocity v by $f = B(v - v_0)$ where f is the effective force on the robot and B is a scalar. Given a nominal commanded velocity v_0^*, the control uncertainty is represented by a cone of velocities (B_{ec} in the figure). The actual commanded velocity v_0 must lie within this cone.

The goal in fig. 17 is to move to the region G. Now, with Coulomb friction, sticking occurs on a surface when the (actual) commanded velocity points into the friction cone. We assume the friction cones are such that sliding occurs (for all possible commanded velocities in B_{ec}) on all surfaces save G, where all velocities stick. We will assume that the planner can monitor position and velocity sensors to determine whether a motion has reached the goal. Velocity sensing is also subject to uncertainty: for an actual velocity v, the sensed velocity lies in some cone B_{ev} of velocities about v.

Now we introduce simple model error. The shape of A and B are known precisely, and the position of A is fixed. However, the position of B, relative to A is not known. B's position is characterized by the distance α. If $\alpha > 0$ the goal is reachable. But if $\alpha = 0$, then the goal vanishes. No plan can be guaranteed to succeed if $\alpha = 0$ is possible. Suppose we allow α to be negative. In this case the blocks meet and fuse. Eventually, for sufficiently negative α, B will emerge on the other side of A. In this case, the goal "reappears," and may be reachable again.[14] Let us assume that α is bounded, and lies in

[14]This model is adopted for the purposes of exposition, not for physical plausibility. It is not hard to model the case where the blocks meet but do not fuse.

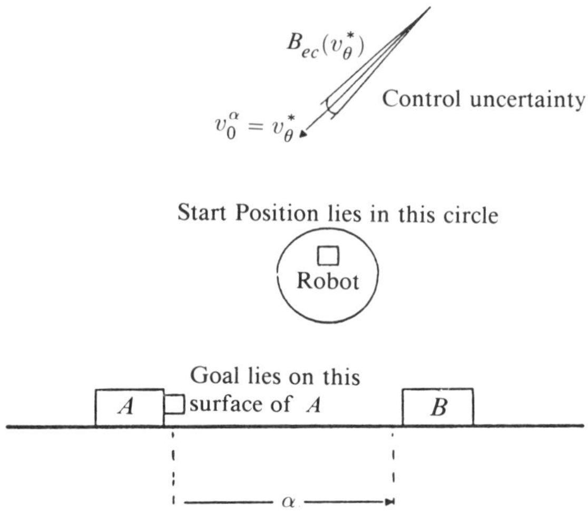

Figure 16: The goal is to bring the robot into contact with the right vertical surface of A. (For example, the "robot" could be a gripper finger). There is position sensing uncertainty, so in the start position the robot is only known to lie within some uncertainty ball. There is also control uncertainty in the commanded velocity to the robot. It is represented as a cone, as shown.

the interval $[-d_0, d_0]$.

Our task is to find a plan that can attain G in the cases where it is recognizably reachable. Such a plan is called a *guaranteed strategy in the presence of model error.* But the plan cannot be guaranteed for the α where the goal vanishes. In these cases we want the plan to signal failure. Loosely speaking, a motion strategy which achieves the goal when it is recognizably reachable and signals failure when it is not is called an *Error Detection and Recovery (EDR) strategy.* Such strategies are more general than guaranteed strategies, in that they allow plans to fail.

To represent model error, we will choose a parameterization of the possible variation in the environment. The degrees of freedom of this parameterization are considered as additional degrees of freedom in the system. For example, in fig. 17, we have the x and y degrees of freedom of the configuration space. In addition, we have the model error parameter α. A coordinate in this space has the form (x, y, α). The space itself is the cartesian product $\Re^2 \times [-d_0, d_0]$. Each *α-slice* of the space for a particular α is a configuration space with the obstacles A and B instantiated at distance α apart. Fig. 17 is such a slice.

More generally, suppose we have a configuration space C for the degrees of freedom of the moving object. Let J be an arbitrary index set which parameterizes the model

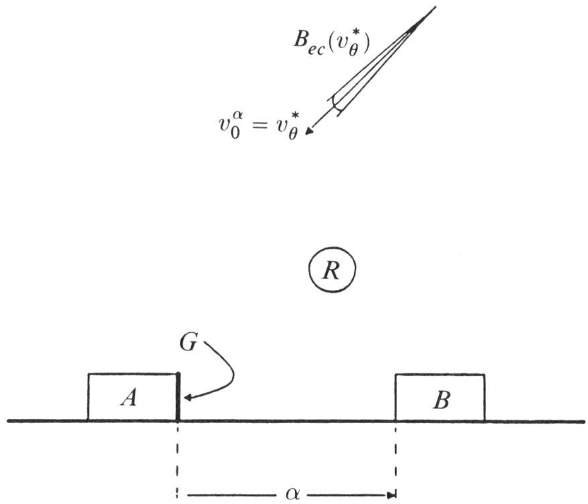

Figure 17: The equivalent problem in configuration space. The blocks A and B, the distance between the blocks α, and the commanded velocity $v_\theta = v_0^*$ with control error cone $B_{ec}(v_0^*)$. The position of A is fixed.

error. (Above, J was $[-d_0, d_0]$). Then the *generalized configuration space* with model error is $C \times J$. One way to think of this construction is to imagine a collection of possible "universes", $\{ C_\alpha \}$ for α in J. Each C_α is a configuration space, containing configuration space obstacles. The ambient space for each C_α is some canonical C. $C \times J$ is simply the natural product representing the ambient space of their disjoint union. There is no constraint that J be finite or even countable.

In fig. 18 we show the generalized configuration space for example (1). Note that the goal in generalized configuration space becomes a 2-dimensional surface, and the obstacles are 3-dimensional polyhedra. Note that the goal surface vanishes where A and B meet.

Given a configuration space corresponding to a physical situation, it is well known how to represent motions, forces, velocities, and so forth in it (eg., see [Arnold]). The representations for classical mechanics exploit the geometry of differentiable manifolds. We must develop a similar representation to plan motions, forces, and velocities in generalized configuration space . We develop the following "axioms" for "physics" in $C \times J$.

1. At execution time, the robot finds itself in a particular slice of $C \times J$, (although it may not know which). Thus we say there is only one "real" universe, α_0 in[15] J. This α_0 is fixed. However, α_0 is not known *a priori*. Thus all motions are confined to a particular (unknown) α_0-slice, such as fig. 17. This is because motions cannot move

[15] α_0 is a point in the multi-dimensional space J.

between universes. In fig. 18, any legal motion in $C \times J$ is everywhere orthogonal to the J-axis and parallel to the x-y plane.

2. Suppose in any α-slice the position sensing uncertainty ball about a given sensed position is some set B_{ep}. The set R in fig. 17 is such a ball. We cannot sense across J: position sensing uncertainty is infinite in the J dimensions.[16] Thus the position sensing uncertainty in $C \times J$ is the cylinder $B_{ep} \times J$. In figs. 17,18, this simply says that x and y are known to some precision, while α is unknown. The initial position in fig. 17 is given by $R \times [-d_0, d_0]$. This cylinder is a 3-dimensional solid, orthogonal to the x-y plane and parallel to the J-axis in fig. 18.

3. Suppose in the configuration space C, the velocity control uncertainty about a given nominal commanded velocity is a cone of velocities B_{ec}. Such a cone is shown in fig. 17. This cone lies in the tangent bundle (or phase-space) TC of C. (Phase space is simply Position-space $\times$ Velocity-space. A point in phase space has the form (x, v), and denotes an instantaneous velocity of v at configuration x). Phase space represents all possible velocities at all points in C. The phase space for $C \times J$ is obtained by indexing TC by J to obtain $TC \times J$. All velocities in generalized configuration space lie in $TC \times J$. For Ex. (1) $TC \times J$ is $\Re^4 \times [-d_0, d_0]$. The generalized velocity uncertainty cones are two-dimensional, parallel to the x-y plane, and orthogonal to the J axis.

4. Generalized damper dynamics extend straight-forwardly to $C \times J$, so motions satisfy $f = B(v - v_0)$ where f, v, and v_0 lie in $TC \times J$. Thus friction cones from configuration space (see [Erdmann]) naturally embed like generalized velocity cones in $TC \times J$.

These axioms give an intuitive description of the physics of $C \times J$. A formal axiomatization is given in [Don 89]. We have captured the physics of $C \times J$ using a set of *generalized uncertainties*, friction, and control characteristics (1-4). These axioms completely characterize the behavior of motions in $C \times J$.

2.4.2 Pushing

By relaxing axiom (1), above, we can consider a generalization of the model error framework, in which pushing motions are permitted, as well as compliant and gross motions. We relax the assumption that motion between universes is impossible, and permit certain motions across J. Consider example (3). Observe that a displacement in J corresponds to a displacement in the position of the block B. Thus a motion in J should correspond to a motion of B. Suppose the robot can change the position of B by pushing on it, that is, by exerting a force on the surface of B. The key point is that pushing operations may

[16] One generalization of the framework would permit and plan for sensing in J. In this case one would employ a bounded sensing uncertainty ball in the J dimensions.

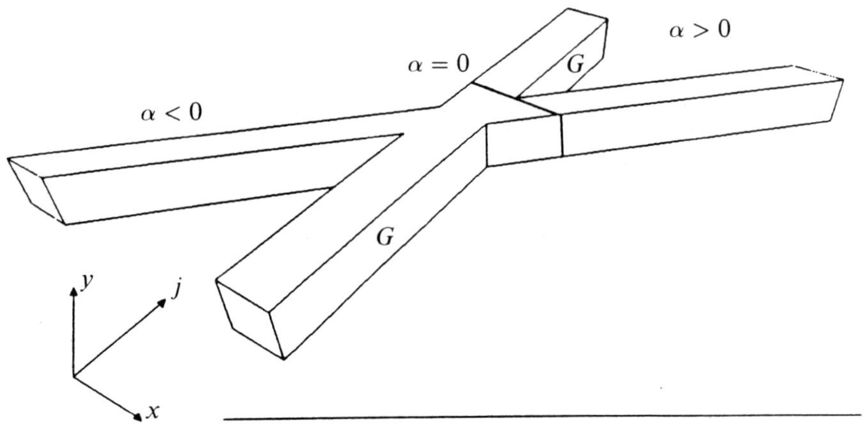

Figure 18: The generalized configuration space obstacles for example (1). The generalized configuration space is three dimensional, having x and y degrees of motion freedom, and an α degree of model error freedom. Legal motions are parallel to the x-y plane, and orthogonal to the J axis.

be modeled by observing that commanded forces to the robot may result in changes in the environment. That is, a commanded force to the robot can result in motion in C (sliding) as well as motion in J (pushing the block). Let us develop this notion further.

Our previous discussion assumed that motion across J was impossible. That is, all motion is confined to one α-slice of generalized configuration space. In example (3), this is equivalent to the axiom that B does not move or deform under an applied force. Such an axiom makes sense for applications where B is indeed immovable, for example, if A and B are machined tabs of a connected metal part. However, suppose that B is a block that can slide on the table. Then an applied force on the surface of the block can cause the block to slide. This corresponds to motion in J. In general, the effect of an applied force will be a motion which slides or sticks on the surface of B, and which causes B to slide or stick on the table. This corresponds to a coupled motion in both C and J, that is, a motion *across α-slices of generalized configuration space.* Such a motion is always tangent to a surface in generalized configuration space.

In [Don89], we generalize the description of the physics of $C \times J$ to permit a rigorous account of such motions. This model can then be employed by an automated planner. *Such a planner can construct motion strategies whose primitives are gross motions, compliant motions, and pushing motions.* This model of pushing is used in the gear-meshing example, where a model error parameter—the orientation of B—can be changed due to pushing.

2.4.3 Guaranteed Plans

Recall that a motion strategy is a commanded velocity v_θ^* together with a *termination predicate* which monitors the sensors and decides when the motion has achieved the goal. Given a goal G in configuration space, we can form its *preimage*. The preimage of G is the region in configuration space from which all motions are guaranteed to move into G in such a way that the entry is recognizable. That is, the preimage is the set of all positions from which all possible trajectories consistent with the control uncertainty are guaranteed to reach G recognizably.

Preimages provide a way to construct guaranteed plans for the situation with no model error. Can preimages and backprojections be generalized to situations with model error? The answer is yes. The generalized control and sensing uncertainties in $C \times J$ are given by the physics axioms above. These uncertainties completely determine how motions in generalized configuration space must behave. We form the backprojection of G under these uncertainties. The trick here is to view the motion planning problem with n degrees of motion freedom and k degrees of model error freedom as a planning problem in an $(n+k)$-dimensional generalized configuration space, endowed with the special physics described above. The physics is characterized precisely by axioms defining certain special sensing and control uncertainties in $C \times J$. The definitions and results for pre-images and backprojections (sec. 2.2) in configuration space generalize *mutatis mutandis* to $C \times J$ endowed with this physics; this i proved in [Don89]. Thus our framework reduces the problem of constructing guaranteed motion strategies with model error to computing preimages in a somewhat more complicated, and higher-dimensional configuration space. For details, see [Don89].

2.5 Error Detection and Recovery

If we were exclusively interested in constructing guaranteed motion strategies in the presence of model error, we would be done defining the framework: having reduced the problem to computing preimages in $C \times J$, we could now turn to the important and difficult problems of computing and constructing $C \times J$, and computing preimages in generalized configuration space.

However, guaranteed strategies do not always exist. In example (1), (figs. 16–18) there is no guaranteed strategy for achieving the goal, since the goal may vanish for some values of α. Because tolerances may cause gross topological changes in configuration space, this problem is particularly prevalent in the presence of model error. In a peg-in-hole problem with model error (fig. 3) the goal may also vanish (the hole may close up) for certain regions in J. More generally, there may be values of α for which the goal may still exist, but it may not be reachable. For example, in a variant of the problem in fig. 6, an obstacle could block the channel to the goal. Then G is non-empty, but also not reachable. Finally, and most generally, there may be values of α for which the goal is reachable but not *recognizably* reachable. In this case we still cannot guarantee plans, since a planner cannot know when they have succeeded.

These problems may occur even in the absence of model error. However, without model error a guaranteed plan is often obtainable by back-chaining and adding more steps to the plan. In the presence of model error this technique frequently fails: in example (1), no chain of recursively-computed preimages can ever cover the start region $R \times J$. The failure is due to the peculiar sensing and control characteristics (1-4) in generalized configuration space .

In response, we will develop Error Detection and Recovery (EDR) strategies. These are characterized as follows:

- An EDR strategy should attain the goal when it is recognizably reachable, and signal failure when it is not.
- It should also permit serendipitous achievement of the goal.
- Furthermore, no motion guaranteed to terminate recognizably in the goal should ever be prematurely terminated as a failure.
- Finally, no motion should be terminated as a failure while there is any chance that it might serendipitously achieve the goal due to fortuitous sensing and control events.

These are called the "EDR Axioms", they will be our guiding principles. We now state how such EDR strategies may be constructed; for proofs and more detail, see [Don89].

For technical reasons (see [Don89]), in the EDR theory, we must construct preimages relative not to the initial set R, but rather to the forward projection $F_\theta(R)$. We will henceforth mercifully abbreviate $P_{F_\theta(R),\theta}$ by P_θ.

Suppose that a planning problem is given with two disjoint geometrical goals, G_1 and G_2. We may insist that the run-time executor be able to terminate the strategy and also be able to disambiguate *which* goal has been reached. In this case, we can construct the *preimage of the distinguishable union of* G_1 *and* G_2, which we write as

$$P_\theta(\{\, G_1, G_2 \,\}).$$

If θ is executed starting in this preimage, then the motion can always be recognizably and distinguishably terminated in either G_1 or G_2.

We can characterize EDR strategies geometrically as follows. Suppose the geometric goal is G. The implicit "meaning" of G is: recognizably achieving G is equivalent to "success." We introduce an "additional" goal-like set H, which is disjoint and distinguishable from G, such that when H is recognizably achieved, then failure of the motion may be signalled. That is, we construct an H such that recognizably achieving H is equivalent to "failure." H is called the *EDR region*. Remarkably, H may be selected such that the EDR

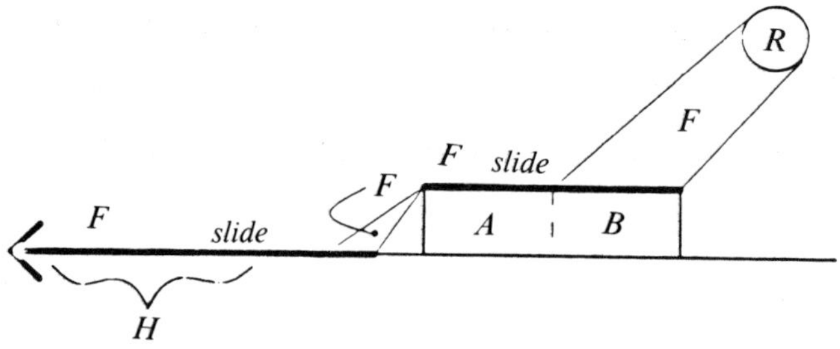

Figure 19: A typical α-slice of the EDR region H, for α small and negative. The goal vanishes in this slice; the dashed line indicates where the goal would be in other slices. Points in H lie within the forward projection (since they are reachable), yet outside the weak-preimage (since the goal is unachievable).

axioms are satisfied. In [Don89], we derive H as follows. Given a motion θ and a start region R, first we define H using reachability constructs only. Then we test whether H and G are distinguishable using sensors (this is the "formal test" alluded to in the prelude). If so, then by the construction of H, we have

$$R \subset P_\theta(\{G, H\}).$$

Furthermore, using H, θ is a one-step EDR strategy satisfying the EDR axioms. Here is an idea of what H is like: In fig. 19 the EDR region H is shown (in position space only) for example (1). Consider H as a two-dimensional region in $C \times J$; just a slice of it is shown in fig. 19. Note that in this example, H only exists in the slices in which G vanishes. Here, given the sensing uncertainty bounds of example (1), the termination predicate can distinguish between G and H based on position sensing, velocity sensing, or elapsed time. Clearly, H satisfies the EDR axioms: the motion is guaranteed to terminate recognizably in G iff the motion began in a universe in which G does not vanish. Otherwise, the motion terminates recognizably in H. In the first case, the termination predicate signals success, in the latter, failure.

Here is how we construct H. Recall the *forward projection* $F_\theta(R)$ of a set R under θ is all positions and velocities that are possibly reachable from R under v_θ^* (subject to control uncertainty). Forward projections only address reachability: the termination predicate is ignored and only the control uncertainty bound and commanded velocity v_θ^* are needed to

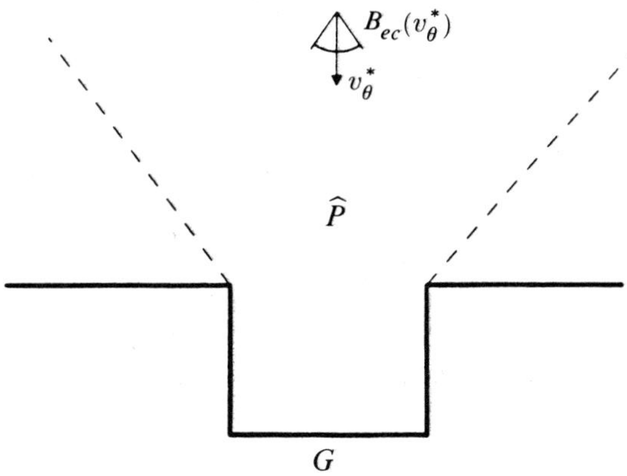

Figure 20: The *weak* preimage of the goal G under v_θ^*. Compare figs. 6 and 12.

specify the forward projection.

So far the preimages we have considered are *strong* preimages, in that *all* possible motions are guaranteed to terminate recognizably in the goal. The *weak* preimage [LMT] (with respect to a commanded velocity) is the set of points which could *possibly* enter the goal recognizably, given fortuitous sensing and control events. See fig. 20. We will use the weak preimage to capture the notion of serendipity in the EDR axioms. The idea is that a motion may be terminated in failure as soon as egress from the weak preimage is recognized. The weak preimage is denoted $\widehat{P}_\theta(G)$.

We define H_0 to be the set difference of the forward projection minus the weak preimage:

$$H_0 = \pi F_\theta(R) - \widehat{P}_\theta(G). \tag{20}$$

Clearly, the motion θ can be terminated as a failure whenever H_0 has been reached, since H_0 is outside the weak preimage; hence the goal cannot be attained under θ from there.

We also define H_s to be all regions where sticking is possible in the weak minus strong preimage:

$$H_s = \{\, x \in \widehat{P}_\theta(G) - P_\theta(G) \mid \text{sticking is possible at } x \,\}.$$

See fig. 21. The motion θ should also be terminated as a failure if sticking occurs in H_s. We will decree that the robot has stuck in H_s if its velocity is zero for some duration,

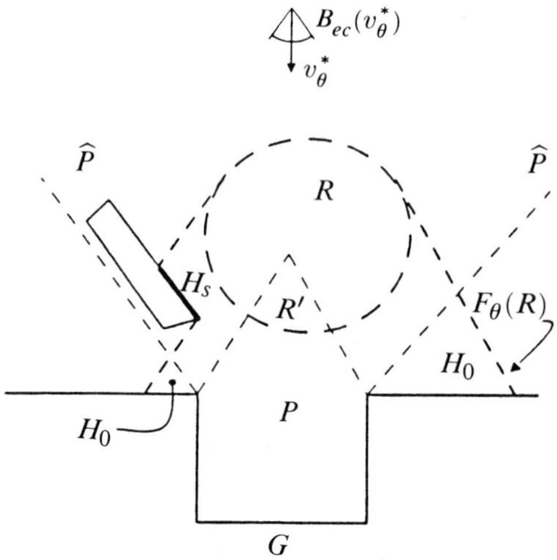

Figure 21: H_0 in eq. (20) is not the entire EDR region. Sticking may occur within the weak preimage in H_s. The EDR region must include H_0 for all possible velocities, and H_s for "sticking velocities."

or "time-out" period.[17] More precisely, we define H to be a set in phase-space. First, note that H contains all phase-space points (x, v) where x is in H_0. Second, H contains points of the the form $(x, 0)$, where x is in H_s. Thus, viewing phase-space as the tangent bundle to generalized configuration space, H contains the "cylinder" $\pi^{-1}(H_0)$ of velocities over H_0, and the "zero section" $Z(H_s)$ of zero velocities over H_s:

$$H = \pi^{-1}(H_0) \cup Z(H_s).$$

This definition of H almost satisfies the EDR axioms—the only tricky point is that we cannot guarantee that after sticking in H_s for a long time, the robot cannot eventually slide into the goal. This may be handled in principle by introducing a time-out period by which the goal must be reached. That is, our definition of H satisfies the EDR axioms if the goal is specified in phase-space-*time* as the product of $\pi^{-1}(G)$ with a compact time interval.

2.5.1 The Preimage Structure of EDR Regions

Consider an equation such as (13) once more, viewing H as an EDR region:

$$R \subset P_{F_\theta(R),\theta}(\{\, G, H \,\}). \tag{21}$$

[17]That is, the termination predicate halts the motion when the velocity is zero for some prespecified duration.

We remarked that in general, solving preimage equations like (13) and (21) for R is like finding the fixed point of a recursive equation. However, in the EDR application, we know R, H, and G, so (21) is a constraint which must be true, rather than an equation to solve. Presumably (21) is easier to check than to solve for R.

2.5.2 Application: Algorithms for EDR Strategies

In [Don89], we showed how to compute H in the domain of planar assemblies with model error. We also showed how to compute whether G and H are distinguishable. This is sufficient to generate one-step EDR strategies. These algorithms have been implemented in LIMITED. At a high level, the one-step algorithm is:

Algorithm Planar One-Step EDR

1. *Generate a commanded velocity* v_θ^*.
2. *Compute the EDR region* H *for* v_θ^*,
3. *Determine whether the EDR region* H *and the goal* G *are distinguishable using sensors. If so, then* v_θ^* *yields a one-step EDR strategy which recognizably terminates in* G *or* H *by monitoring position and force sensors.*
4. *Let* $\text{push}_\theta(G)$ *and* $\text{push}_\theta(H)$ *denote the* sticking push-forwards. *They are the set of obstacle edges within* G *and* H*, resp., on which sticking can occur under* v_θ^*. *Determine whether these regions are distinguishable using sensors. If so, then* v_θ^* *yields a one-step EDR strategy which recognizably terminates when sticking is detected.*

Here is how LIMITED decides the question, "Are G and H distinguishable using sensors?"

H and G are distinguishable using position sensing alone if their convolutions (Minkowski sums) by the position sensing error ball B_{ep} do not intersect.

Each obstacle edge of H and G has an associated configuration space friction cone. Two edges are distinguishable using force sensing if the convolutions of their friction cones by the force sensing uncertainty B_{ev} have a trivial intersection.

Similarly, the set of possible sensed reaction forces at an obstacle vertex w of G or H may be found by taking the direct sum of the friction cones of the edges cobounding w, and convolving by B_{ev}. Again, a vertex of H and a vertex (or edge) of G are distinguishable using force sensing if their associated cones of sensed reaction forces have a trivial intersection.

The procedure also works for determining the distinguishability of the push-forwards. Note that the procedure is correct for linear edges, where position- and force-sensing are separable, because the set of possible reaction forces is constant along an edge. For the general case, see [Don89].

2.6 Advanced Topics

Now, we have only addressed the planning of single-step EDR strategies. In order to plan multi-step EDR strategies, the techniques of sec. 2.3 may be employed. Multi-step strategy synthesis is complex, and the theory extends rather far beyond this paper.

We note that, in principle, having reduced both model error and EDR to essentially "preimage-theoretic" equations, multi-step strategies could be synthesized by solving these preimage equations. While this is proved or at least implicit in previous work [LMT,E,Ma2,D], it is far from obvious; furthermore, there are almost no published examples of such strategies. For this reason we presented a worked-out example of a motion plan using preimages in sec. 2.3. We could easily derive EDR as well as guaranteed strategies by solving the preimage equations [Don90].

Preimages are a key underlying tool for the geometric EDR theory, and the LMT framework is in some sense a "universal" method for synthesizing multi-step strategies. However, the technique of solving the preimage equations is not computational. For this reason, we have introduced a construction called the *push-forward* [Don89]. Roughly speaking, the push-forward is that subset of the forward projection where the motion can terminate. Since push-forwards address termination whereas forward projections do not, we may regard them as "dual" to preimages. That is, push-forwards are to forward projections as preimages are to backprojections. Second, the push-forward permits us to develop rather simple algorithms for planning multi-step strategies. These algorithms have been implemented in LIMITED. While the push-forward method for multi-step strategy synthesis is algorithmic, it is less general than the full preimage method (solving the preimage equations). We characterize the loss of power in push-forward algorithms in [Don89].

In [Don89], we presented two EDR plans generated by LIMITED. These were a peg-in-hole insertion strategy with model error, and a gear-meshing plan. Both were two-step plans. The peg-in-hole plan used push-forward techniques. The gear plan used a seemingly unrelated technique called *failure mode analysis*. However, there exists a view of multi-step strategies which essentially unifies all these techniques. This is called the "weak" EDR theory in [Don89]. The motivation behind this theory is that when a motion terminates ambiguously, a subsequent motion may be synthesized which disambiguates the success or failure of the first. Oddly enough, it is not necessary for either motion individually to satisfy the EDR axioms of sec. 2.5. However, when taken together, the two-motion plan can often be considered "equivalent" to a one-step EDR strategy.

The weak EDR theory effectively defines some laws of "composition" that permit two single-step plans to be concatenated into a two-step plan satisfying the EDR axioms. Hence it is often possible to construct multi-step plans that are EDR plans "globally" although not "locally". That is, considered as entire plans, they satisfy the EDR axioms; this is the "global" condition. However, "locally" they are not EDR plans, in that no single step is an EDR strategy. The key to pasting together non-EDR plans to make a global EDR strategy lies in defining certain local "niceness" conditions for how plans must mesh. These are called the *linking conditions.*

As the notation suggests, it is possible to formalize our view of "motions as mappings"—this notion is implicit in the term "preimage." To develop this viewpoint, one considers motions as a certain class of morphisms between distinguishable unions in the powerset of the tangent bundle to generalized configuration space . An EDR theory, then, is a covariant functor associated with a family of quotient maps of the form

$$\pi : TC \times J \rightarrow \{ P, \hat{P} - P, \tilde{F} - \hat{P} \}.$$

One focus for future work would be to push such a functorial viewpoint; any category-theoretic formulation of this flavor is still operational in nature and not yet fully abstract.

Acknowledgments: I am very grateful to Randy Brost for providing me with figures 8-10. Also, thanks to Tomás Lozano-Pérez, Matt Mason, Russ Taylor, Mike Erdmann, and John Canny for sharing with me the exciting research I survey in this paper.

3 References

Abraham, R. and Marsden, J. *Foundations of Mechanics,* Benjamin/Cummings, London (1978).

Arnold, V. I. *Mathematical Methods of Classical Mechanics,* Springer-Verlag, New York (1978).

[BK] Bajaj, C. and M. Kim, "Compliant Motion Planning with Geometric Models", *Proc. ACM Symposium on Computational Geometry*, Waterloo, 1987.

Brady, M. et. al. (eds). *Robot Motion: Planning and Control.,* Cambridge, Mass.: MIT Press. (1982).

Briggs, Amy *An Efficient Algorithm for One-Step Compliant Motion Planning with Uncertainty,* 5th ACM Symposium on Computational Geometry, Saarbrucken, Germany. (1989).

Brost, R., "Automatic Grasp Planning in the Presence of Uncertainty", *IEEE International Conference on Robotics and Automation*, San Francisco, April, 1986.

Brost, R. *A State/Action Space Approach to Planning Robot Actions,* Forthcoming Ph.D. Thesis, Computer Science Dept., CMU (to appear).

Brost, R. *Computing Metric and Topological Properties of Confuguration Space Obstacles,* IEEE Int. Conference on Robotics and Automation, Scottsdale, AZ (1989).

Buckley, S. J. *Planning and Teaching Compliant Motion Strategies,* Ph.D. Thesis. Massachusetts Institute of Technology, Department of Electrical Engineering and Computer Science, 1987. Also MIT-AI-TR-936 (1987).

Burridge, R., Rajan, V. T., and Schwartz, J. T. *The Peg-In-Hole Problem: Statics and Dynamics of Nearly Rigid Bodies in Frictional Contact,* IEEE ICRA, Raleigh, NC (1983).

[C1] Canny, J.F. *The Complexity of Robot Motion Planning,* Ph.D. Thesis, MIT Department of Electrical Engineering and Computer Science (1987).

[Can89a] John Canny. *On computability of fine motion plans.* , IEEE ICRA, Scottsdale, AZ, (1989)

[CD] Canny, J.F. and Donald, B. R. *Simplified Voronoi Diagrams,* Discrete and Computational Geometry **3** (3) pp. 219-236. (1988).

[CR] Canny, J., and J. Reif, "New Lower Bound Techniques for Robot Motion Planning Problems", *FOCS* (1987).

[D] Donald, B. R. *Robot Motion Planning with Uncertainty in the Geometric Models of the Robot and Environment: A Formal Framework for Error Detection and Recovery,* IEEE International Conference on Robotics and Automation, San Francisco, April (1986a).

Donald, B. R. *A Search Algorithm for Motion Planning with Six Degrees of Freedom,* Artificial Intelligence, 31 (**3**) (1987a).

Donald, B. R. *Error Detection and Recovery for Robot Motion Planning with Uncertainty,* Ph.D. Thesis, Dept. EECS, MIT Artificial Intelligence Laboratory, MIT-AI-TR 982 (1987b).

Donald, B. R. *The Complexity of Planar Compliant Motion Planning with Uncertainty,* Proc. ACM Symposium on Computational Geometry, Urbana, IL June, 1988. *To appear in* Algorithmica. (1988a).

[D] Donald, B. R. *A Geometric Approach to Error Detection and Recovery for Robot Motion Planning with Uncertainty,* Artificial Intelligence **37** (1-3), Dec. 1988. pp. 223-271. (1988b).

Donald, B. R. *Planning Multi-Step Error Detection and Recovery Strategies,* Proc. IEEE International Conference on Robotics and Automation, Philadelphia, PA April, 1988. (1988c).

Donald, B. R. *Error Detection and Recovery in Robotics,* Lecture Notes in Computer Science, Vol. 336, Springer-Verlag, New York (1989). (1989).

Donald, B. R. *Planning Multi-Step Error Detection and Recovery Strategies,* International Journal of Robotics Research, **3** (1), March, (In press). (1990).

Draper Laboratories, Fourth Annual Seminar on Robotics and Advanced Assembly Systems, Cambridge, Massachusetts, November, 1983.

Dufay, B., and J. Latombe, "An Approach to Automatic Robot Programming Based on Inductive Learning", in Brady, M., and R. Paul, *Robotics Research: The First International Symposium,* MIT Press, 1984.

Durrant-Whyte, H. *Concerning Uncertain Geometry in Robotics,* Intl. Workshop on Geometric Reasoning, Oxford, England, June (1986).

[E] **Erdmann, M.** *Using Backprojections for Fine Motion Planning with Uncertainty,* IJRR Vol. 5 no. 1 (1986).

Erdmann, M. *On Motion Planning With Uncertainty,* MIT AI Lab, MIT-AI-TR 810 (1984).

Erdmann, M. *On Probabilistic Strategies for Robot Tasks,* Ph.D. thesis, MIT Department of EECS, MIT A.I. Lab, Cambridge (1989)..

[EM] **Erdmann, M., and M. Mason**, "An Exploration of Sensorless Manipulation", *IEEE International Conference on Robotics and Automation*, San Francisco, April, 1986.

[FHS] **Friedman, J., J. Hershberger and J. Snoeyink** *Compliant Motion in a Simple Polygon,* 5th ACM Symposium on Computational Geometry, Saarbrucken, Germany. (1989).

[GM89] **Goldberg, K., and Mason, M.** *Bayesian Grasping,* Submitted to the IEEE Int'l. COnf. on Robotics and Automation, 1990 (1989).

Hopcroft, J. E., Schwartz, J. T., and Sharir, M. 1984 On the Complexity of Motion Planning for Multiple Independent Objects; *PSPACE*-Hardness of the "Warehouseman's Problem." *International Journal of Robotics Research.* **3**(4):76–88.

Hopcroft J., and Wilfong G., "Motion of Objects in Contact," Int. Jour. Robotics Res. vol 4, no. 4, (1986).

Inoue, H., "Force Feedback in Precise Assembly Tasks", MIT Artificial Intelligence Laboratory, AIM-308, August, 1974.

J. Jennings, B. Donald, and D. Campbell *Towards Experimental Verification of an Automated Compliant Motion Planner based on a Geometric Theory of Error Detection and Recovery,* Proc. IEEE International Conference on Robotics and Automation, Scotsdale, Arizona (1989).

[LLS] **Latombe, J.C., A. Lazanas, and S. Shekhar** *Motion Planning with Uncertainty: Practical Computation of Non-Maximal Preimages,* IEEE/RSJ Intl. Workshop on Intelligent Robots and Systems, Sept. 4-6 (1989).

Laugier, C., "A Program for Automatic Grasping of Objects with a Robot Arm", *Eleventh Symposium of Industrial Robots*, Japan Society of Biomechanisms and Japan Industrial Robot Association, 1981.

Lieberman, L., and M. Wesley, "AUTOPASS: An Automatic Programming System for Computer Controlled Mechanical Assembly", *IBM Journal of Research Development*, Vol. 21, No. 4, 1977, pp. 321-333.

Lozano-Pérez, T., "The Design of a Mechanical Assembly System", S.M. dissertation, MIT Department of Electrical Engineering and Computer Science, also AI-TR-397, MIT Artificial Intelligence Laboratory, 1976.

Lozano–Pérez, T. *Automatic Planning of Manipulator Transfer Movements,* IEEE Trans. on Systems, Man and Cybernetics (SMC–11):681–698 (1981).

[LoP] Lozano–Pérez, T. *Spatial Planning: A Configuration Space Approach,* IEEE Trans. on Computers (C–32):108–120 (1983a).

Lozano-Pérez, T., "Robot Programming", *IEEE Proceedings*, 1983b.

Lozano-Pérez, "A Simple Motion Planning Algorithm for General Robot Manipulators," in Proceedings of Fifth National Conference for the American Association of Artificial Intelligence, Philadelphia, 1986, pp. 626–631.

[LMT] Lozano-Pérez, T., Mason, M. T., and Taylor, R. H. *Automatic Synthesis of Fine-Motion Strategies for Robots,* Int. J. of Robotics Research, Vol 3, no. 1 (1984).

Lozano-Pérez, T., and Wesley, M. A. *An algorithm for planning collision-free paths among polyhedral obstacles,* Communications of the ACM (22):560-570 (1979).

Lumelsky, V. J. *Continuous Motion Planning in Unknown Environment for a 3D Cartesian Robot Arm.* , *Proceedings of the 1986 IEEE International Conference on Robotics and Automation*, pp. 1569–1574. (April 7–10, San Francisco, Calif.) (1986).

Mason, M.T. *Compliance and force control for computer controlled manipulators,* IEEE Trans. on Systems, Man and Cybernetics (SMC-11):418-432 (1981).

Mason, M. T. 1986. Mechanics and Planning of Manipulator Pushing Operations. *International Journal of Robotics Research* **5**(3).

[Ma2] Mason, M. T. *Automatic Planning of Fine Motions: Correctness and Completeness,* 1984 IEEE International Conference on Robotics, Atlanta Ga. (1984).

Natarajan, B. K. 1986 (Oct. 27–29, Toronto, Ontario). An Algorithmic Approach to the Automated Design of Parts Orienters. *Proceedings of the 27th Annual IEEE Symposium on Foundations of Computer Science*, pp. 132–142.

Ohwovoriole, M., and B. Roth, "A Theory of Parts Mating For Assembly Automation", *Proceedings of the Robot and Man Symposium 81*, Warsaw, Poland, September 1981.

Peshkin, M., "Planning Robotic Manipulation Strategies for Sliding Objects", Ph.D. dissertation, Department of Physics, Carnegie-Mellon University, 1986.

Raibert, M., and J. Craig, "Hybrid Position/Force Control of Manipulators", *Journal of Dynamic Systems, Measurement, and Control*, No. 102, June, 1981, pp. 126-133.

Schwartz J., Hopcroft J., and Sharir M., "Planning, Geometry and Complexity of Robot Motion Planning", Albex Publishing Co., New Jersey, (1987).

Schwartz J. and Yap C. K., "Advances in Robotics," Lawrence Erlbaum associates, Hillside New Jersey, (1986).

Simunovic, S. N., "An Information Approach to Parts Mating", Ph.D. dissertation, Department of Electrical Engineering, Massachusetts Institute of Technology, 1979.

Taylor, R. H. *The Synthesis of Manipulator Control Programs from Task-level Specifications,* Stanford Artificial Intelligence Laboratory, AIM-282, July (1976).

Turk, M., "A Fine-Motion Planning Algorithm", *SPIE Conference on Intelligent Robots and Computer Vision*, Cambridge, Massachusetts, September, 1985.

Udupa S., "Collision Detection and Avoidance in Computer Controlled Manipulators", Proc. 5th Int. Joint. Conf. on Art. Intell., Mass. Inst. Tech. (1977) pp 737-748.

Valade, J., "Automatic Generation of Trajectories for Assembly Tasks", *Sixth European Conference on Artificial Intelligence*, Pisa, Italy, September, 1984.

[Whi76] Whitney, D., "Force Feedback Control of Manipulator Fine Motions", *Journal of Dynamic Systems, Measurement, and Control*, June, 1977, pp. 91-97.

[Whi82] Whitney, D., "Quasi-Static Assembly of Compliantly Supported Rigid Parts", *Journal of Dynamic Systems, Measurement, and Control*, Vol. 104, March 1982.

Department of Computer Science, Cornell University, Ithaca, NY 14853.

Proceedings of Symposia in Applied Mathematics
Volume **41**, 1990

Formal Languages for Motion Description and Map Making[1]

R. W. Brockett

1. Review of Formal Models

There are a number of formal models used for defining the word computable. Perhaps the simplest of these is that of a finite state machine.

Let U be a finite set and define $U^\star$ to be the set of all finite strings of elements of U. We think of $U^\star$ as being a monoid with concatenation being the binary operation. Since we impose no bound on the length of the strings in $U^\star$, this set is infinite. Let Y be a second finite set. We consider a function $f : U^\star \to Y$ and ask about the difficulty of computing (evaluating) f. A sample question, vaguely phrased, is this: let $U = Y = \{0,1\}$ and let f_1 be the function which assigns to an element of $U^\star$ the mod 2 sum of its entries; let f_2 be the function which assigns the value 1 to a string consisting of an arbitrary number of ones followed by exactly the same number of zeros and assigns the value zero to all other strings. Which of these two functions is "easier" to compute?

By a finite automaton one understands a five tuple $(U, X, Y, \delta, \gamma)$ where U, X and Y are finite sets, $\delta : X \times U \to X$, $\gamma : X \to Y$. The operation of the automaton is defined by

$$x(k+1) = \delta[x(k), u(k)];\ \gamma[x(k)] = y(k)$$

which together with a specification of an initial state defines a mapping from $U^\star$ into Y. An automaton with a specified initial state is called an *initialized automaton.*

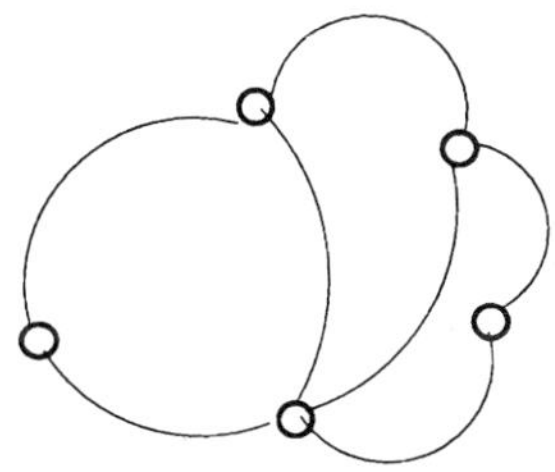

Figure 1. A five state finite automaton with some transitions missing.

We visualize the action of the automaton with the help of a diagram. The elements of X are represented by nodes. The branches connecting the nodes indicate the action of the automaton on receipt of a particular input from U. The idea can be seen in figure 1. This together with a definition of γ defines the automaton.

1980 *Mathematics Subject Classification* (1985 *Revision*). Primary 68E05, 68F05, 93C99, 93E10.

[1]This work was supported in part by the U.S. Army Research Office under grant DAAL03-86-K-0171 and in part by the National Science Foundation under grant CDR-85-00108.

0160-7634/90 $1.00 + $.25 per page

Suppose that E and F are subsets of the free monoid $U^\star$. By $E \cup F$ and $E \cdot F$ we understand the subsets

$$\begin{aligned} E \cup F &= \{u | u \in E \text{ or } u \in F\} \\ E \cdot F &= \{u | u = ef \text{ with } e \in E\ ,\ f \in F\} \end{aligned}$$

we also define $E^\star$ to be the set of all finite strings of elements of E

$$E^\star = \{u | u = e_i, e_{i2} \dots e_{in}\ ;\ e_i \in E\}$$

In other words $E^\star$ is the smallest submonoid of $U^\star$ which contains E. One says that a subset L of $U^\star$ is *regular* if it is finite or if it can be generated from a finite subset of $U^\star$ by a finite number of applications of the operations $\cup$, $\cdot$, and $\star$.

A subset L of $U^\star$ is a *finite state language* if there exists an initialized finite state machine $(U, X, Y, \delta, \gamma)$ with $Y = \{0, 1\}$ such that for any input string in L the output is 1 and for any input string not in L the output is 0.

2. Kinematic Machines and Motion Languages

In the first section, we saw that a finite automaton is a simple model of computation, suitable for the definition of some classes of languages. Our goal now is to structure our notion of motion accordingly—in particular, to identify formal languages that describe certain classes of motion and to construct machines that model the computation, or generation, of such classes of motion. Given such specially tailored linguistic and computational devices, we should be better able to evaluate motion programming languages with respect to their expressiveness, efficiency, complexity, etc.

The first task at hand is to explain what we mean by a *formal motion language.* A motion language, simply stated, is a set of symbolic strings that represent idealized motions. Stated with more precision, a motion language L is a subset of a free monoid on a given motion alphabet Σ, that is, $L \subseteq \Sigma^\star$. Symbols of alphabet Σ represent segments of motion, having specific durations and various other characteristics. Although this portrayal of a motion language appears to fit well with the traditional view of a formal languages, this is not quite the standard fare. While one normally does not think of the symbols of an alphabet in formal language theory in terms of their effect on a particular machine, one must do so in the case of motion languages. Consequently we define motion languages only in the context of an associated physical model, which we call a *kinematic machine.*

In a very broad sense, a kinematic machine can be thought of as a continuous analog of a finite automaton. Whereas finite automata are governed by difference equations of the form

$$x_{k+1} = \delta(x_k, u_k);\quad y_k = \gamma(x_k)$$

kinematic machines are governed by differential equations of the form

$$\dot{x} = \Sigma g_i(x) u_i;\quad y = h(x)$$

Recall that we defined a finite automaton as a quintuple $(U, X, Y, \delta, \gamma)$, where U is a finite input alphabet, X is a finite set (the machine states), Y is a finite output alphabet, $\delta : X \times U \to X$ is a transition function that maps current states and inputs into next states, and $\gamma : X \to Y$ is a map that associates each machine state with an output symbol. In a like manner, we may denote a kinematic machine as a quintuple (U, X, Y, G, h), where U is an input (control) space, X is an actuator coordinate space (state space), Y is an output space, G is a matrix whose entries may depend on x, and $h : X \to Y$ transforms actuator coordinates to output coordinates. These characterizations of machine behavior mark their essential similarities and differences. Important points to note are (1) the state space for a finite automaton is a discrete, finite set whereas the state space for a kinematic machine is a continuum, (2) both finite automata and kinematic machines prescribe a transformation from the state space of the machine state X to the output space Y, and (3) an initial state condition (x_0 or $x(0)$) and an input sequence $u_0 \ldots u_n \in U^*$ determine the state evolution for both finite automata and kinematic machines.

Henceforth, we adopt a vector notation and consider kinematic machines to be models of the form

$$\dot{x} = G(x)u; \quad y = h(x)$$

where u, x, and y are functions of time. The input u is subject to some regularity conditions, say u belongs to the set of all piecewise continuous functions. x has its range in n-dimensional actuator coordinate space X. y has its range in p-dimensional output space Y (Y typically has position coordinates and force coordinates). G is an n by m matrix that depends only on x. $h : X \to Y$ is an actuator-to-output coordinate transformation.

The atoms of our motion language include triples of the form (u, k, T) whose significance we now explain. If at time t_0 the machine receives an input (u, k, T), the state x evolves according to

$$\dot{x} = G(x)(u + k(y)); \quad y = h(x); \quad 0 \le t \le T.$$

If at time t_0 the machine receives an input string $(u_1, k_1, T_1) \ldots (u_i, k_i, T_i)$, then x evolves according to

$$\begin{aligned} \dot{x} &= G(x)(u_1 + k_1(y)); & t_0 \le t < t_0 + T_1 \\ &\vdots & \vdots \\ \dot{x} &= G(x)(u_i + k_i(y)); & t_0 + T_1 + \ldots + T_{i-1} \le t \le t_0 + T_1 + \ldots + T_i \end{aligned}$$

where $k : Y \to U$ is a feedback function belonging to a function space K. With this encapsulation, we see that provisions have been made for the generation of motion (open-loop and closed-loop motion are both special cases). See figure 2.

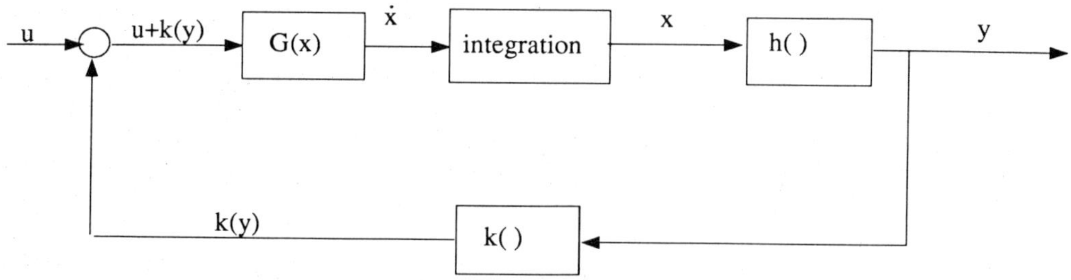

Figure 2: Feedback loop associated with a kinematic machine.

For the sake of simplicity, and to give one concrete example of a motion language, we define the language $M1$ to be the set of strings of triples (u_i, k_i, T_i), where u_i are constant, k_i are linear functions from Y to U (i.e., identifiable with matrices), and T_i are positive real numbers.

Example: The following is a depiction of a specific kinematic machine which has been built to test these theories. In figure 3, a robotic device is shown with two 2-link fingers and pressure sensors mounted at the tips of each. Here actuator coordinates are θ_1, θ_2, θ_3, and θ_4, and output coordinates are y_1 through y_6. Then the kinematic machine is governed by these equations.

$$\begin{bmatrix} \dot\theta_1 \\ \dot\theta_2 \\ \dot\theta_3 \\ \dot\theta_4 \end{bmatrix} = \begin{bmatrix} u_1 \\ u_2 \\ u_3 \\ u_4 \end{bmatrix}; \quad \begin{bmatrix} y_1 \\ y_2 \\ y_3 \\ y_4 \\ y_5 \\ y_6 \end{bmatrix} = h\left(\begin{bmatrix} \theta_1 \\ \theta_2 \\ \theta_3 \\ \theta_4 \end{bmatrix}\right) = \begin{bmatrix} 5\cos\theta_1 + 3\cos(\theta_1 - \theta_3) \\ 5\sin\theta_1 + 3\sin(\theta_1 - \theta_3) \\ 7 - 5\cos\theta_2 + 3\cos(\theta_4 - \theta_2) \\ 5\sin\theta_2 + 3\sin(\theta_4 - \theta_2) \\ \text{pressure}_1(\theta_1, \theta_2, \theta_3, \theta_4) \\ \text{pressure}_2(\theta_1, \theta_2, \theta_3, \theta_4) \end{bmatrix}.$$

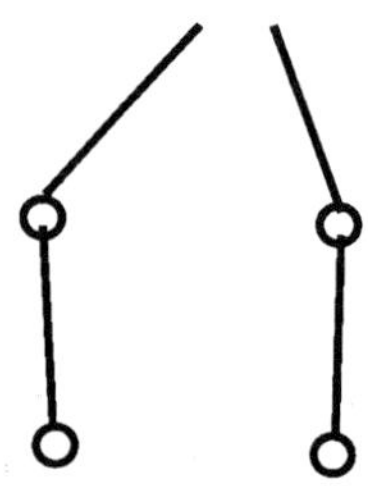

Figure 3. A four dgree of freedom kinematic machine.

3. World Models

We want a model of the world which will serve to describe the force/displacement relationships associated with robots interacting with objects. We do not want a detailed model but we need one which is qualitatively correct. We begin with a few points.

(i) It is to be expected that both the robot and the objects it manipulates are compliant to some extent.

(ii) Forces and torques are vector quantities. In many situations these quantities can be thought of as being derived from a single function, a potential, by way of

$$f = -\nabla V$$

(iii) In some situations one achieves an economy of thought by transferring properties of a moving object to the fixed objects which populate the workspace – in effect giving them a shadow and shrinking the moving object to a point. We will use this kind of thinking to model compliance.

(iv) We must make a distinction between world coordinates and robot coordinates. We must not assume that the robot has a perfect description of the world.

Our model of the world is that it is free space populated with a number of rigid objects B_i which are, to begin with, fixed in space. We assume that we have chosen a euclidean coordinate system for the world with coordinates (w_1, w_2, w_3). Neither the locations of the objects nor their shapes are known with precision. Associated with each *fixed object* is a potential function $V_i : \mathbf{E}^3 - B_i \to \mathbf{R}$ which maps the exterior of the ith object into the real line according to the formula

$$V_i(w) = \begin{cases} \frac{1}{d} - \frac{d}{\epsilon^2} + \frac{2}{\epsilon^2} & \text{if } w \text{ is of distance} < d \text{ to } B_i \\ 0 & \text{if } w \text{ is of distance } d > 0 \text{ from } B_i \end{cases}$$

(Although V_i is not defined on the interior of B_i, one might think informally that it takes on the value $+\infty$ there.)

When the robot is in position y, it feels a force due to object B_i which is

$$f_i = -\nabla V_i|_y$$

We also will want to consider world models with *movable rigid objects*. Again, in this case we must be careful not to assume that the location or the shape of the object is known exactly. In this case we can associate with each movable object B_i a potential as above, and also a convex subset K of the space of forces and torques such that if a net applied force/torque belongs to this set, support and frictional forces will keep B_i fixed. Thus the description of a movable rigid object is more complex than that of a fixed rigid object by virtue of the need for K.

Example: Consider a world model which consists of one movable object in $\mathbf{E}^2$. The object is a disk and of radius $2 \pm .1$. The associated K is, forces related to the center,

$$\begin{array}{c} -1 \leq f_x \leq 1 \\ -1 \leq f_y \leq 1 \\ -1/2 \leq T \leq l/2 \end{array}$$

Exercise:

Consider a disk of radius one centered at $w_1 = w_2 = 0$ and its associated potential $V(w_1, w_2)$ given by the formula of this section. Compute the associated Hessian, $H(w_1, w_2)$. Establish a tolerance μ such that if $(w_1 - z_1)^2 + (w_2 - z_2)^2 < \mu$, then the eigenvalues of

$$M(w_1, w_2, z_1, z_2) = -H(z_1, z_2) \cdot H(w_1, w_2)$$

have negative real parts.

4. Stability Issues; The Potential Method

We begin by defining three machines. These examples are point machines in that they occupy no volume. First consider a kinematic machine which operates in one dimension and senses its position and force

$$\dot{x} = u \; ; \; y_1 = x \; ; \; y_2 = f \qquad \text{(MACHINE I)}$$

with f being the force felt by the machine. Secondly, consider a machine which operates in the plane and senses position and force

$$\begin{array}{l} \dot{x}_1 = u_1 \; ; \; y_1 = x_1 \; ; \; y_3 = f_1 \\ \dot{x}_2 = u_2 \; ; \; y_2 = x_2 \; ; \; y_4 = f_2 \end{array} \qquad \text{(MACHINE II)}$$

Here f_i is the force in the x_i direction. Finally, we consider a three degree of freedom machine which operates in the plane but which admits rotation as well as translation. It senses position and angle as well as force and torque

$$\begin{array}{l} \dot{x}_1 = u_1 \; ; \; y_1 = x_1 \; ; \; y_4 = f_1 \\ \dot{x}_2 = u_2 \; ; \; y_2 = x_2 \; ; \; y_5 = f_2 \\ \dot{x}_3 = u_3 \; ; \; y_3 = x_3 \; ; \; y_6 = \tau \end{array} \qquad \text{(MACHINE III)}$$

Now f_i is the force in the x_i direction and τ is the torque felt by the machine.

Suppose that we have a world model of the form discussed in the previous section. Consider the response of Machine I to the language atom (u, k, T) (1,[0,2],10). According to our conventions, this gives an equation of motion

$$\dot{x} = 1 - 2\nabla V \qquad 0 \leq t \leq 10$$

We cannot, of course, say what the trajectory will be until we specify a world model. Say that the world model has an object of length 3 centered at $w = 9$, say that ϵ is .5 and say that $x(0) = -1$. What will happen on $t \in [0, 10)$ is sketched in figure 4. Without making any difficult calculations we can assert that $x(10)$ is between 8.0 and 9.5, that it is in contact with the object and that neither of these predictions is altered if the position of the center of the interval is changed from 9 to $9 + \alpha$ as long as $|\alpha| < 2$. Likewise T can be changed to any number greater than 8.5 and the conclusions are not altered. *This is an extremely robust situation.*

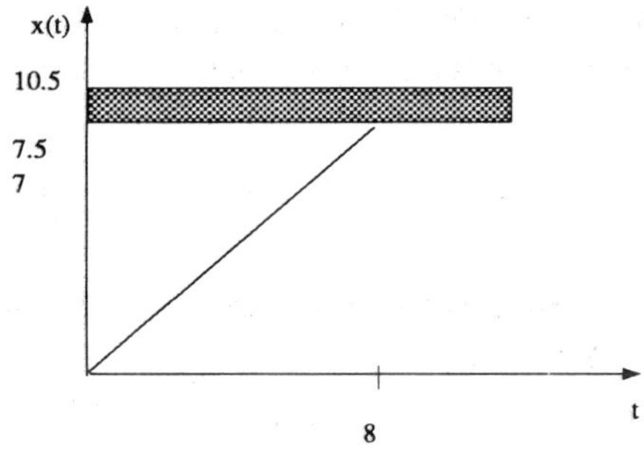

Figure 4: The motion which results from (u, k, T).

It is likewise true that the behavior is not strongly dependent on the world model. The $1/d - 1/\epsilon$ potential could be altered to be any differentiable strictly monotonic function which goes from zero at .5 units from the edge to infinity at the edge. We leave the proof of this as an exercise.

In the two dimensional situation matters become more delicate. Consider figure 5.

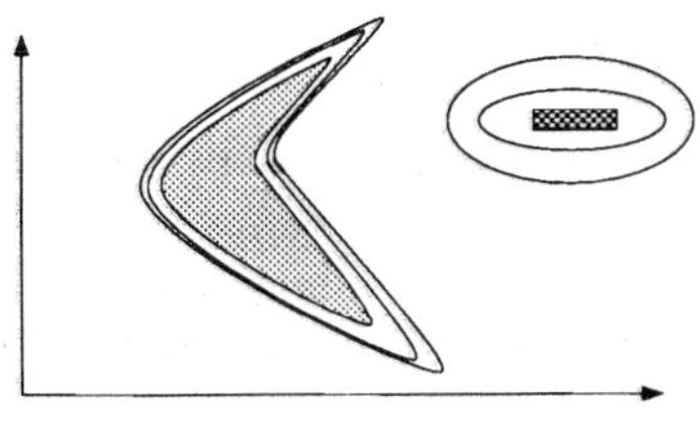

Figure 5: Object and Equipotential lines in $\mathbf{E}^2$

If we turn Machine II loose in this environment, what will happen? In order to analyze this situation it is useful to associate a potential with u and to analyze the sum of $V(B_i)$ and the u-dependent potential. This works as follows. Notice that if we think of V as being expressed as a function of x then

$$\dot{x} = -\nabla V(B) + u = -\nabla(V(B) - x^T u)$$

Thus we can think of this rule as bringing x to the bottom of the valley of the function $V(B) - x^T u)$.

We still have k at our disposal, however, and it can be used to advantage. What we see is that in the case of a convex object such as (b) above, the function $V(B) - x^T u$ has a saddle point and not a local minimum near the point of contact. Notice that near this saddle point

$$V_u = V(B) - x^T u \simeq q + \frac{1}{2}(\delta x)' H(\delta x) \quad ; \quad \delta x = (x - x_0)$$

with $H = H^T$ and H having one positive and one negative eigenvalue. Near this saddle point, for a given $k(y) = Ky$ we have

$$\delta \dot{x} = -KH\delta x$$

Notice that if we would choose K to be H, then

$$\delta \dot{x} = -H^2 \delta x$$

and regardless of the eigenvalues of H, those of H^2 are nonnegative (zero only if H has a zero eigenvalue). Thus this equation is asymptotically stable.

We do not know the location of B exactly so we do not know V_u exactly so we do not know H exactly. Even so, except for the degenerate case (i.e., the case where H has a zero eigenvalue) we can say that the solutions of

$$\dot{x} = -(H + \delta H)\nabla V_u$$

will approach an equilibrium point.

Definition: We will call the choice (u, k, T) where $k(y) = Hy$ *descent programming*.

Although we will not go into the three dimensional version of these problems here, it should be clear that descent programming will work in three dimensions as well.

5. Changes of Coordinates

The essence of graceful motion is the smooth coordination of the available degrees of freedom. From one point of view the basic idea of a kinematic machine requires nothing further; better coordination is just a matter of choosing better sequences of (u_i, k_i, T_i). On the other hand, computational considerations and programming efficiency suggest that one should incorporate into the language some tools involving coordinate change and trajectory smoothing to ease the programmers task.

In the basic definition of a kinematic machine there is a confounding of the effects of k and u in the sense that a given $\dot{x}$ can be the result of a large k and small u or a large u and small k. To disambiguate this we need to reflect on the different mechanisms which are to be used to generate kx and u and to consider the different response speeds.

In this chapter, in fact, let us write the differential equations which result from applying $(u, [k_1, k_2], T)$ to Machine II as follows

$$\dot{x} = k_1(x - x_0) + k_2 \nabla V + (u + k_1 x_0)$$

We may think of the equation as follows:

$$\text{Velocity} = \text{posture term} + \text{navigation term} + \text{planning term}$$

In some cases it may be useful to split x into two terms according to

$$x = \begin{bmatrix} x_{\text{pose}} \\ x_{\text{plan}} \end{bmatrix} \ ; \ x_{\text{plan}} \in Ker\ k_1 \ ; \ x_{\text{pose}} \in Range\ k_1$$

In the language of psychology the term involving x_{pose} is proprioceptive feedback and k_1 would be said to define a conditioned reflex. It is to be observed that kinematic machines must have a certain level of complexity if posture is to be a worthwhile idea.

Because the u and k elements in (u, k, T) are taken to be constants the equations of motion for the machine

$$\dot{x} = k_1 x + k_2 \nabla V + u$$

generate trajectories of a particular type. For example, if we let k_1 and k_2 be zero then $x(t)$ will take the form $x(t) = x_0 + u \cdot t$, a straight line in x-space. On the other hand, if we change coordinates, letting x be replaced by $z = g(x)$, then

$$\dot{z} = (\partial g / \partial x)\, \dot{x} = (\partial g / \partial x) u$$

If we were to trade the input u for $v = G(g^{-1}(z))u$, then we get

$$\dot{z} = v$$

for the kinematics of the machine in the new coordinate system. In this case a constant input gives rise to a straight line in z-coordinates. It is desirable, at the same time, to change the definition of the force in such a way as to maintain the energy-theoretic meaning of the inner product between input and force output. This means letting $y = \nabla V$ with the gradient being with respect to z. Interpreting triples (v, k, T) as inputs to this machine effectively adapts the programming problem to z-coordinates.

More formally, if g is a differentiable one to one mapping with a differentiable inverse, we will allow it to be an input to our machine, interpreting it as an instruction to change coordinates before executing the next triple (u, k, T). Thus the admissible input strings now include sequences including subsequences of the form

$$\ldots\ (u_1, k_1, T_1) g_1 (u_2, k_3, T_3)(u_3, k_3, T_3) g_2\ \ldots$$

Of course it takes additional computation to support such an addition to the language. The essential additional computation is the evaluation of the Jacobian G of g and the mapping of u by G.

6. Recording and Mapping

In this section we are going to assume that all machines measure their position in addition to whatever else they measure. Consider a kinematic machine

$$\dot{x} = u \quad ; \quad y = f$$

So far we have in the motion language elements (u, k, T) and change of coordinate transformations g. We now add the final ingredient, R, T, which asks the computational structure which supports the machine to store the measurement record for the past T units of time and a variance matrix $\sum$ which provides some indication as to how reliable x can be expected to be.

The effect of reaching the ith occurrence of an R, T will be to record the measurement record as the ith entry in a file.

We now turn to a discussion of how the record function might be used in map making. Consider the results of a robot which has made measurements about the world. After a certain period, the map file will contain some locations and some variances associated with those locations. As distance information is recorded, we create a directed graph. Our notation for this is $G = (V, E)$, where V is the set of vertices of G and E is the set of edges of G. Vertices of G are labelled with locations, and edges of G are labelled with means and variances of distances between locations. If we are making a two-dimensional map, these means are 3-vectors.

Initially, the graph G is just a root vertex v_0; it has no edges. The root vertex is labelled with an initial location l_0. The graph G is then updated incrementally. Each time a measurement is taken, the current distance is computed from the most recent pair of location measurements. Figure 6 presents a typical picture. Associated with the entire set of measurements is a Gaussian density with a certain mean, d, and certain variance, Σ_d. By introducing an orientation on the edges we can associate a vector distance with each edge. If the map is planar, we are modeling the d's as two component vectors. The first component being the x component of the distance, the second being the y component.

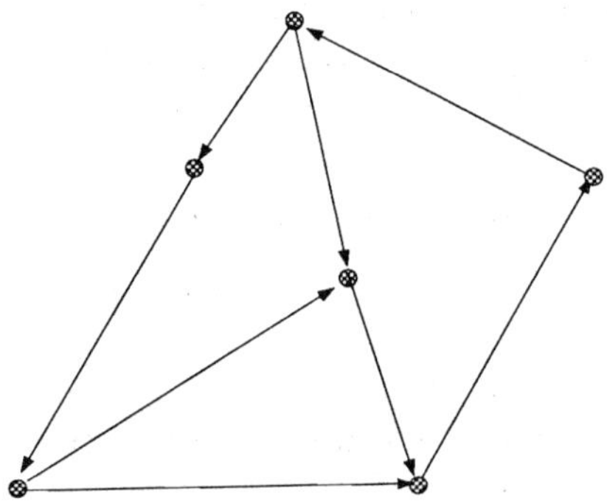

Figure 6. A graph representing the contents of a data file.

The measurements $d_1, \ldots, d_n$ are not, however, independent since geometry imposes some constraints. These constraints can be expressed easily by selecting a *maximal tree* in the graph; i.e., a subgraph which contains no loops and which cannot be enlarged without losing this property. (The maximal tree may not be unique.) We denote by d_t the vector of these distances along the branches of the maximal tree. Using the fact that the sum of the d_i's around any closed loop is zero, we can find an incidence matrix A such that

$$Ad_t = d$$

Notice that the entries of A need only take on the values (+1,0,-1). If there are k branches in the maximal tree and n branches altogether, then A is n by k and of rank k.

At a certain point in time, we may wish to analyze the map. For this purpose, we identify a spanning tree (or maximal tree) T. Any one of the remaining edges forms a cycle when added to the tree. The a priori Gaussian density on the edge measurement is, say

$$p(d) = N\, e^{-\frac{1}{2}(d-\bar{d})^T\, \Sigma^{-1}(d-\bar{d})}$$

However, when we impose the constraint that d can be expressed as Ad_t the density becomes

$$p(d) = N_1\, e^{-\frac{1}{2}(Ad_t-\bar{d})^T\, \Sigma^{-1}(Ad_t-\bar{d})}$$

By maximizing this we see that the most likely tree-distance vector d_t is given by

$$d_t = (A^T\Sigma^{-1}A)^{-1}\, A^T\Sigma^{-1}\bar{d}$$

This formula can be interpreted as giving a *rationalization* of the raw measurements represented by $\bar{d}$. The effect of this process is to shift the "best guess" of d from $\bar{d}$ to

$$\hat{d} = A(A^T\Sigma^{-1}A)^{-1}\, A^T\Sigma^{-1}\bar{d}$$

The correction being

$$(\hat{d} - \bar{d}) = (A(A^T\Sigma^{-1}a)^{-1}A^T\Sigma - I)\bar{d}$$

Now suppose that we have a graph G which contains information about means and variances of distances. For any k, the expected value for the observed distance at the kth edge in G is $\bar{d}_k$

$$\bar{d}_k = d^t \bar{d}_T$$

and the variance associated with this estimate of the distance along the kth edge is

$$\sigma_k = c^t \Sigma_T c = c^t (A^t \Sigma^{-1} A)^{-1} c \, .$$

Assume that we measure the length of the kth edge with an associated variance σ_m. The posterior estimate for the distance measurement at the kth edge of G is a linear combination of the prior estimate $\bar{d}_k$ and the actual distance measurement d_m.

$$\hat{d}_k = \frac{\sigma_m}{\sigma_m + \sigma_k} \bar{d}_k + \frac{\sigma_k}{\sigma_m + \sigma_k} d_m$$

This equation shows how the kth edge gets updated. If the kth edge was not in the tree, then there are cycles in G that depend on this edge, and the edges of these cycles should reflect the information gained from the measurement d_m. This is done by updating the edges of T, as the edges of G depend linearly on those of T.

If the observation d_m was made on a new edge of G, then d and Σ get augmented.

$$d = \begin{bmatrix} d \\ d_m \end{bmatrix} = \begin{bmatrix} A \\ c^t \end{bmatrix} d_T$$

Because d_m is independent of past measurements, covariances are zero in the new Σ.

$$\Sigma = \begin{bmatrix} \Sigma & 0 \\ 0 & \sigma_m \end{bmatrix}$$

The new tree variance Σ_T retains the same dimensions as before.

$$\Sigma_T = \left(\begin{bmatrix} A^t & c \end{bmatrix} \begin{bmatrix} \Sigma & 0 \\ 0 & \sigma_m \end{bmatrix}^{-1} \begin{bmatrix} A \\ c^t \end{bmatrix} \right)^{-1} = \left(\Sigma_T^{-1} + c \sigma_m^{-1} c^t \right)^{-1}$$

By the matrix inversion lemma, this equation can be expressed as a formula which requires only one inverse.

$$\Sigma_T = \Sigma_T - \Sigma_T c (\sigma_m + \sigma_k)^{-1} c^t \Sigma_T$$

Finally, the new estimate for tree distances is a linear combination of the old estimate and the measurement.

$$\hat{d}_T = \left(\Sigma_T^{-1} + c(\sigma_m - c^t \Sigma_T c)^{-1} c^t \right)^{-1} \left(\Sigma_T^{-1} \bar{d}_T + c(\sigma_m - c^t \Sigma_T c)^{-1} c^t d_m \right)$$

This is the conditional expectation of d_T given d_m. It can be simplified.

$$\hat{d}_T = (\Sigma_T - \Sigma_T \sigma_k \Sigma_T)(\Sigma_T^{-1} \bar{d}_T + c(\sigma_m - c^t \Sigma_T c)^{-1} c^t d_m)$$

Reference

[1] R. W. Brockett, "On the Computer Control of Movement," *Proceedings of the IEEE Conference on Robotics and Automation*, 1988.

[2] C. Blume and W. Jacob, *Programming languages for Industrial Robotics*, Springer-Verloy, Berlin, 1986.

[3] Randell Smith, Matthew Smith and Peter Cheeseman, "A Stochastic Map for Uncertain Spatial Relationships", *IEEE Conference on Robotics and Automation* IEEE, New York, 1987.

Index